An Introduction to Combinatorics

An Introduction to Combinatorics

Alan Slomson

School of Mathematics
University of Leeds

CHAPMAN AND HALL
LONDON • NEW YORK • TOKYO • MELBOURNE • MADRAS

UK	Chapman and Hall, 2–6 Boundary Row, London SE1 8HN
USA	Chapman and Hall, 29 West 35th Street, New York NY10001
JAPAN	Chapman and Hall Japan, Thomson Publishing Japan, Hirakawacho Nemoto Building, 7F, 1-7-11 Hirakawa-cho, Chiyoda-ku, Tokyo 102
AUSTRALIA	Chapman and Hall Australia, Thomas Nelson Australia, 102 Dodds Street, South Melbourne, Victoria 3205
INDIA	Chapman and Hall India, R. Seshadri, 32 Second Main Road, CIT East, Madras 600 035

First edition 1991

Typeset in 10/12 Times by
Keytec Typesetting Ltd, Bridport, Dorset
Printed in Great Britain by
T. J. Press (Padstow) Ltd, Padstow, Cornwall

ISBN 0 412 35360 1 (HB)
0 412 35370 9 (PB)

British Library Cataloguing in Publication Data
Slomson, Alan
Counting.
1. Combinatorial analysis
I. Title
511.6

ISBN 0–412–35360–1
ISBN 0–412–35370–9 pbk

Library of Congress Cataloging-in-Publication Data
Slomson, A. B.
An introduction to combinatorics/Alan Slomson.—1st ed.
p. cm.—(Chapman and Hall mathematics series)
Includes bibliographical references and index.
ISBN 0–412–35360–1 (U.S. : hard).—ISBN 0–412–35370–9 (U.S. : pbk.)
1. Combinatorial analysis. I. Title. II. Series.
QA164.S57 1991
511′.6—dc20 90–24182
CIP

To Gina, in appreciation for having persuaded me to buy a word processor and with apologies for using it to write a book without a single murder in it.

Contents

Introduction

WHY COMBINATORICS?

I am greatly attracted by combinatorics because it is a branch of mathematics that deals with concrete, easy to understand, problems. Mathematics is a problem-solving activity, and the ultimate source of all mathematics is the external, non-mathematical world. Mathematical concepts are developed to help us tackle these problems. The abstract mathematical ideas that we use soon assume a life of their own, and generate further problems, but these are more technical problems whose connection with the external world is often remote.

This leads to a problem in the teaching and learning of mathematics. It is generally thought that it is best to begin with the most basic ideas and then gradually work your way up to more and more complicated mathematics. This seems more sensible than being thrown in at the deep end and hoping you will learn to swim before you drown. However, this logical approach to learning mathematics obscures the historical reasons why a particular mathematical idea was developed. This means that it is often difficult for the student to understand the real point of the subject.

I do not in any way intend to belittle technical abstract mathematics. The solution of technical mathematical problems can make a big contribution to solving the mathematical problems arising from the real world which ultimately gave rise to them. The difficulty is that pure mathematics often seems very remote. The concepts are difficult to assimilate, and students who do not understand the historical context in which the theory grew up, or the problems which the theory solves, do not know what to make of it.

Combinatorics is different.

The starting point in combinatorics usually consists of problems which are easy to understand. They are concrete problems which can be understood by those who do not know any technical mathematics. Many of the chapters in this book begin with a problem of this kind, and then go on to develop the mathematics needed to solve it.

Since the time of Isaac Newton the emphasis in applied mathematics has been on continuously varying processes, modelled by the mathematical continuum, and using methods derived from the differential and integral calculus. In contrast, combinatorics concerns itself with *finite* collections of *discrete* objects. With the growth in digital devices, especially digital computers, discrete mathematics has become more and more important. The flavour of discrete, combinatorial mathematics is very different from that of continuous mathematics. However, the two areas interact. In this book I do not hesitate to use methods drawn from continuous mathematics where these are appropriate. Indeed, one of the aims of the book is to show students how the abstract pure mathematics which they learn, be it convergence of series from analysis, or cosets from group theory, or whatever, is relevant to the solution of problems drawn from the real world.

COUNTING PROBLEMS

The hardest choices in writing a book of this kind lie in deciding what to put in and what to leave out. Combinatorics covers a wide range of different problems, so one possibility for an introductory text would be to include a small sample from each of the main areas. I prefer a book with a good plot and so I have decided instead to stick to a common theme, that of *counting problems*.

Counting problems arise when the combinatorial problem is to count the number of different arrangements of collections of objects of a particular kind. Such counting problems arise frequently when we want to calculate probabilities and so they are of wider application than might appear at first sight. The book begins with counting problems of a very straightforward kind and ends with the complicated problem of counting the number of different graphs with a given number of vertices. There are a few digressions and the occasional sub-plot, but most of the topics have been chosen to fit into the general theme.

Some counting problems are very easy, others are extremely difficult. I have tried to make the book intelligible, without creating the impression that combinatorics is not a serious part of mathematics. The hope is that most readers will find a majority of the book straightforward but that they will also meet some challenging problems. Inevitably the choice of topics reflects my own interests and taste, which I hope the reader will share. There is a reading list at the end of the book for those who wish to explore both more complicated counting problems and other areas of combinatorics.

WHAT YOU NEED TO KNOW

You will be glad to know that for most of the book all you are assumed to know is a little elementary algebra. In a few places, and especially in the proof of Stirling's asymptotic formula for $n!$ in Chapter 4, it is assumed that the reader is familiar with some elementary mathematical analysis. I have deliberately included this non-combinatorial chapter in order to try to show one significant use of analysis, a branch of mathematics whose motivation often seems unclear. The parts of the book which make use of ideas from calculus or analysis can safely be omitted, but, of course, I would not have included them if I had not hoped you would read them.

Most of the notation I use is standard, or is introduced as I come to it. I end this section with a short list of some notation which may not be familiar to everyone. First, if A and B are sets, then $A \backslash B$ is the set of elements which are in A but not in B. That is

$$A \backslash B = \{x : x \in A \text{ and } x \notin B\}.$$

Second, I use the 'arrow' notation for functions. For example, the function which squares each number will be written as $x \mapsto x^2$. If $f : D \to C$, that is, f is a function which maps the set D to the set C, then D is called the **domain** of f and C is called the **codomain** of f.

Third, the number of elements in a set X is written as $\#(X)$.

Fourth, the symbols $\mathbb{N}$, $\mathbb{N}^+$, $\mathbb{Z}$, $\mathbb{Q}$, $\mathbb{R}$, $\mathbb{C}$, are used for the following sets of numbers:

$\mathbb{N}$ is the set of **natural** numbers, that is, $\mathbb{N} = \{0, 1, 2, \ldots\}$.
$\mathbb{N}^+$ is the set of **positive integers**, that is, $\mathbb{N}^+ = \{1, 2, 3, \ldots\}$.
$\mathbb{Z}$ is the set of **integers**, that is, $\mathbb{Z} = \{\ldots -2, -1, 0, 1, 2, \ldots\}$.
$\mathbb{Q}$ is the set of **rational numbers**.
$\mathbb{R}$ is the set of **real numbers**.
$\mathbb{C}$ is the set of **complex numbers**.

Note that for me the set of natural numbers includes the number 0. This is a matter which divides mathematicians. You will find that in many books the natural numbers are defined so as to exclude 0, so that $\mathbb{N}$ coincides with the set of positive integers. Of course, whether or not 0 is a natural number, is a matter of convention. Whenever natural numbers are used for counting it is sensible to include 0 so that we have a number to represent the number of elements in an empty set.

Finally, I use $\text{int}(x)$ to mean the integer part of x, for example, $\text{int}(2.27) = 2$. $\ln(x)$ is the natural logarithm of x.

ARE YOU SITTING COMFORTABLY?

Once upon a time there was a programme on the radio for young children called *Listen with Mother*. (In those days it was assumed that it would be the mother who would be at home with the children.) In the first programme in 1950 the storyteller, Julia Lang, introduced the story she was about to tell by saying 'Are you sitting comfortably? Then we'll begin'. Apparently this introduction was not planned, but it caught on, and was used regularly until the programme ceased in 1982.*

When it comes to reading mathematics, however, this is not an appropriate beginning. A mathematics book cannot be read like a novel, sitting in a comfortable chair, with a glass by your side. Mathematics books need to be worked at. You need to be sitting at a table or a desk, with pencil and paper, both to work through the theory and to tackle the problems. A good guide is the amount of time it takes you to read the book. A novel can be read at a rate of about 60 pages an hour, whereas when it comes to many mathematics books you are doing well if you can read five pages an hour. (It follows that, even at 12 times the price, a mathematics book is good value for money!)

Since the approach of the book is to begin with problems and to use them to lead to the theory, there are a number of problems whose solutions are given in the text. The chapters of the book are divided up into sections, and most of these sections end with a set of problems, labelled 'Exercises', some of a routine nature, others more testing.

The answer to a counting problem is usually a number, often a very large number. You are encouraged to do the calculations, and not just look up the answers. These days, with very cheap and powerful calculators, and fairly cheap micro-computers, calculations are very much easier than they used to be. If you can write computer programs, do try to write programs to do the calculations for you when this is appropriate. If you are lucky you will have access to one of the symbolic manipulation programs which now even take the hard work out of elementary algebra and calculus. If you do not have access to any electronic devices, do not be discouraged. All the calculations in this book were first done by hand, though most have been checked by machine.

Solutions to the Exercises are found at the end of the book, and are written out in rather more detail than is usual in books of this kind. This is intended to be helpful, but it will not achieve its purpose of enabling you to learn combinatorics, if you do not make a serious attempt at these problems before turning to the solutions. The book ends with Supplementary Exercises, covering all the chapters, for which no solutions are provided.

*See Nigel Rees, *Sayings of the Century*, London, 1984.

Experience tells me that it is unlikely that all my solutions are completely accurate. Apart from typing errors and minor oversights, there may even be some howlers. I should be pleased to hear from any reader who detects any errors in the solutions, or elsewhere in this book, and indeed from anyone who would like to make any comments of any kind. Please write to me at the School of Mathematics, The University, Leeds LS2 9JT, England.

ACKNOWLEDGEMENTS

None of the mathematics in this book is new. Much of it has been known for a very long time. I have included a few historical references, but I have not made any serious attempt to give a precise history of the topics I have tried to expound. The reading list at the end of the book includes those works from which I have learned most.

The content of this book is based on a series of lectures which I have given to students at the University of Leeds over a period of years. I am grateful to my colleagues in Leeds who first allowed me to teach an area of mathematics in which I am not an expert, and to the students whose comments have helped shape the course.

I have also worked for a number of years for the Open University, mainly as a part-time tutor, but also as a member of one course team. The Open University provides courses for home-based students who receive some tutorial support, but mostly have to study in isolation. I greatly respect the success the Open University has had in teaching mathematics at a distance, and I hope some of their skill has rubbed off on me, so that this book will be accessible to students who study on their own, as well as to those who have the advantage of communal learning. In particular, students of the Open University course M203, *Introduction to Pure Mathematics*, will notice that my treatment of groups has been influenced by the M203 units on this topic.

1

Permutations and combinations

1.1 INTRODUCTION

The main aim of this chapter is to introduce you to the idea of combinatorial thinking. You will probably already have met most of the facts about permutations and combinations that are mentioned, but the combinatorial approach to them may be new to you. For example, in section 1.2 you will find a combinatorial proof of the Binomial Theorem, for which you may previously have only seen an algebraic proof.

In section 1.4 we show how the ideas we have introduced can be used to tackle a variety of probability problems, especially those arising in card games.

Our standard approach is to introduce ideas through problems. You are encouraged to try to solve these problems for yourself before reading the solutions given here.

1.2 PERMUTATIONS

We begin with some very simple problems, but the idea behind their solutions is of fundamental importance in many counting problems.

Problem 1

A café has the following menu:

Tomato soup
Fruit Juice

Lamb chops
Baked cod
Nut roll

Apple pie
Strawberry ice

How many different three course meals could you order?

Solution
You have two choices for the first course, and whichever choice you make, you have a further three choices for the second course. This makes $2 \times 3 = 6$ possibilities for the first two courses:

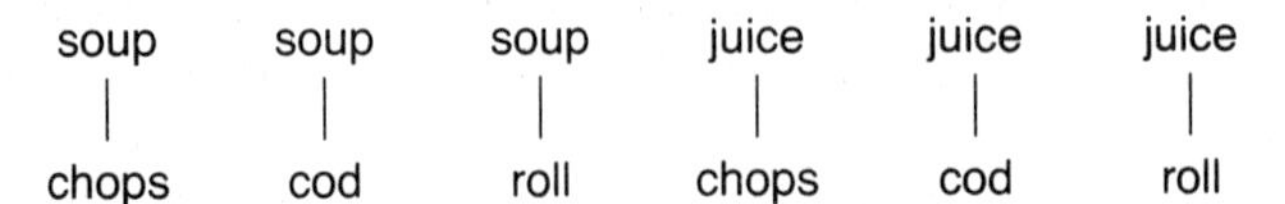

In each of these six cases you have two choices for the third course, making 6×2 possibilities altogether. We can set them out in the following diagram which makes it clear why the number of cases multiplies at each stage, and why the final answer is the product of the number of choices at each stage:

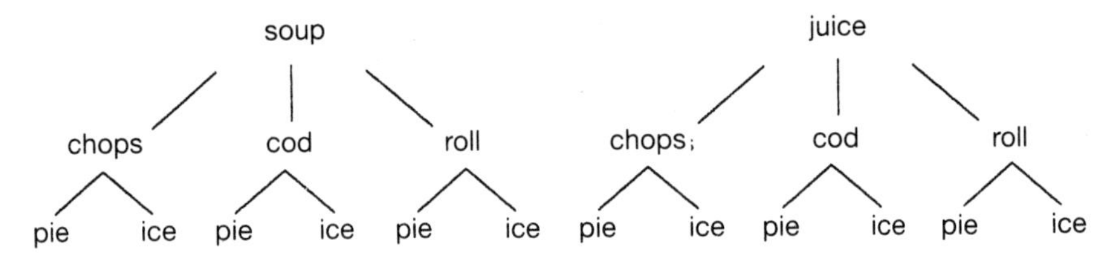

Thus we obtain $2 \times 3 \times 2 = 12$ as the total number of possible meals.

Problem 2
In a race with 20 horses, in how many ways can the first three places be filled?

Solution
There are 20 horses that can come first. Whichever horse comes first, there are 19 horses left that can come second. So there are $20 \times 19 = 380$ ways in which the first two places can be filled. In each of these 380 cases there are 18 horses which can come third. So there are $380 \times 18 = 20 \times 19 \times 18 = 6840$ ways in which the first three positions can be filled.

There is one difference between the situations in these two problems. In Problem 1 the choice of the first course did not affect the choice of the second course. Whether you chose to begin your meal with the soup or the fruit juice, you still had the same three choices, lamb chops, or baked cod or nut roll, for your second course. Similarly, whatever your choice of the first two courses, you had the same choice, the apple pie or the strawberry ice, for your third course.

In Problem 2 the horse that wins the race cannot come second. So the possibilities for which horse comes second vary according to which horse wins the race. However, the *number* of possibilities remains the same. Whichever horse wins, there are 19 horses that can come second, though which particular 19 horses these are varies according to who the winner is. Thus it was correct to multiply 20 by 19 to count the number of ways in which the first two places can be filled. Likewise, the possibilities for the third horse in the race vary according to which two horses come first and second, but, whichever these two horses are, there always remain 18 horses that can come third. So the product $20 \times 19 \times 18$ does indeed count the number of ways the first three positions in the race can be filled.

The multiplication principle we have used in these two problems is sufficiently important to be worth stating as a separate principle.

Theorem 1.1 *(The Principle of Multiplication of Choices)*
If there are k successive choices to be made, and, for $1 \leqslant i \leqslant k$, the ith choice can be made in n_i ways, then the total number of ways of making these choices is

$$n_1 \times n_2 \times \ldots \times n_k$$

that is

$$\prod_{i=1}^{k} n_i.$$

Problem 2 has another feature which is common to problems of this type. The successive choices were all being made from the same fixed set, in this case the set of 20 horses taking part in the race. So the number of horses left to choose from goes down by 1 at each successive choice. In the notation of Theorem 1.1,

$$n_{i+1} = n_i - 1, \qquad \text{for } 1 \leqslant i < n.$$

We introduce some special terminology and notation to describe this type of situation. We call a choice of r objects *in order* from a set of n objects, a **permutation** of r objects from n. We let $P(n, r)$ be the number of different permutations of r objects from n. For this to make sense we need to have $r \leqslant n$. An extension of the method used in Problem 2 yields the general formula for $P(n, r)$.

Theorem 1.2
For all $r, n \in \mathbb{N}$, with $r \leqslant n$,

$$P(n, r) = n(n-1)(n-2) \ldots (n-r+1),$$

that is

$$P(n, r) = \frac{n!}{(n-r)!}.$$

Proof

When we choose r objects in order from n objects, the first object can be chosen in n ways, leaving $n-1$ choices for the second object, and so on. After $r-1$ objects have been chosen there remain $n-(r-1) = n-r+1$ objects from which to make the rth choice. Hence, by the Principle of Multiplication of Choices, the total number of ways of making these choices is

$$n \times (n-1) \times \ldots \times (n-r+1).$$

This number can be rewritten as

$$\frac{n(n-1)\ldots(n-r+1)(n-r)(n-r-1)\ldots 2 \times 1}{(n-r)(n-r-1)\ldots 2 \times 1}.$$

and, using the factorial notation, we can write this expression succinctly as

$$\frac{n!}{(n-r)!}$$

Arranging a set of n objects in order amounts to the same thing as first deciding which object is to be first, then which is to be second, and so on. Thus it is the same as choosing n objects in order from a set of n objects: that is, it is a permutation of those objects. So, by Theorem 1.2, the number of ways of doing this is

$$P(n, n) = \frac{n!}{0!} = n!$$

So we have:

Theorem 1.3
The number of permutations of n objects is $n!$

The values of $n!$ grow very fast. Even for quite small values of n, $n!$ is very large. For example, $10! = 3\,628\,800$, and $100!$ is larger than 10^{157}. In Chapter 4 we describe Stirling's formula, which enables us to obtain very rapidly good approximations to the value of $n!$ for large values of n.

All the examples in this section involve the choice of objects *in order*. In section 1.3 we turn our attention to cases where the order is not relevant.

Exercises

1.2.1. Car licence plates in Britain currently have the form

F123 MUG

that is, a letter, followed by a one-, two- or three-digit number, followed by a further three letters. How many different licence plates of this kind are there?

1.2.2. In three races there are ten, eight and six horses running, respectively. You win a jackpot prize if you correctly predict the first three horses, in the right order, in each race. How many different predictions can be made?

1.2.3. How many sequences are there of n digits in which no two consecutive digits are the same?

1.3 COMBINATIONS

We are now interested in counting the number of ways of choosing objects from a set when the order of selection does not matter. We tackle this problem by relating it to the problem of counting permutations, which we have already solved. (Reducing a new problem to a case which has already been solved is a common mathematical technique. It is said that a mathematician who has learnt how to make a cup of tea starting with an empty kettle, will, when given a full kettle and asked to make tea, first empty the kettle so as to reduce the problem to one that has already been solved.) A couple of examples will make the line of approach clear.

Problem 3

A doubles team is to be selected from a squad of six tennis players. How many different teams can be selected?

Solution

We have seen that we can choose two players, in order, from a squad of six players in $P(6, 2) = 6 \times 5 = 30$ ways. This counts each team of two players twice, since, for example, choosing first Pat and then Chris leads to the same team as first choosing Chris and then Pat. Since we get 30 when we count each team twice, the number of different doubles teams is $30/2 = 15$. We have to divide by 2 because a particular team of two players can be chosen, in order, in $2! = 2$ different ways.

The technique we have used in this problem is used again, not only in the immediately following problems, but in many other counting problems. We count the number of arrangements of a particular kind by counting them in such a way that each of the arrangements is counted more than once. We then adjust our answer to take account of this duplicate counting.

Problem 4

How many different poker hands of five cards can be drawn from a pack of 52 cards?

Solution

We can choose five cards, in order, from a pack of 52 cards in $P(52, 5) = 52!/47!$ different ways. But the order in which the cards are dealt does not matter. The same hand of five cards can be ordered in 5! ways, and so can be chosen, in order, in this number of ways. Thus $P(52, 5)$ gives the number of 5 card hands when each hand is counted 5! times. Hence the number of different poker hands is

$$\frac{P(52, 5)}{5!} = \frac{52!}{47!5!} = \frac{52 \times 51 \times 50 \times 49 \times 48}{5 \times 4 \times 3 \times 2 \times 1} = 2\,598\,960.$$

We can now generalize the method we used in these last two problems. We call a selection of r objects from n objects, when the order does not matter, a **combination** of r objects from n. We use the notation $C(n, r)$ for the number of different combinations of r objects from n.

Notice that the mathematical usage of the words *permutation* and *combination* is the opposite way round from their normal usage in everyday life. We are using *permutation* when the order of the elements is important. In football pools permutations or 'perms' are selections of football matches in which the order does not matter. We are using *combination* when the order of the objects does not matter, but in a combination lock, the order of the numbers is important if you want the lock to open. It is no good having the right set of a numbers if they are not in the right order. It is a mystery to me as to why the standard mathematical usage of *permutation* and *combination* does not follow ordinary usage.

The method we use in Problems 3 and 4 leads us to the general formula for $C(n, r)$.

Theorem 1.4

For all $r, n \in \mathbb{N}$, with $r \leqslant n$,

$$C(n, r) = \frac{n!}{(n - r)!r!}.$$

Proof

We have already discovered that the number of ways in which r objects can be chosen, in order, from n objects is given by $P(n, r) = n!/(n - r)!$ A set of r objects can be ordered in $r!$ different ways. Thus $P(n, r)$ gives the number of r-element subsets of n objects, when each r-element subset is counted $r!$ times. Hence the number of different r-element subsets is

$$\frac{P(n, r)}{r!} = \frac{n!}{(n - r)!r!}.$$

The numbers $C(n, r)$ are very well known. They are usually called *binomial coefficients*. The reason for this name will shortly be explained. A more common notation for the binomial coefficient, which we have written $C(n, r)$ is $\binom{n}{r}$, and you will also find other notations such as ${}_nC_r$ and C_r^n in other books. We have chosen to use $C(n, r)$ mainly because it is easier to print, but also because it fits in with the standard mathematical notation for functions of two variables, $f(x, y)$, and also with the notation for two-dimensional arrays in most programming languages.

The formula for $C(n, r)$ given by Theorem 1.4 can be used to give algebraic proofs of the chief properties of the binomial coefficients. We prefer, however, to emphasize the combinatorial meaning of these numbers, and hence to give combinatorial proofs whenever this is convenient. In line with this approach, we have given a combinatorial definition of the number $C(n, r)$. The alternative would be to define $C(n, r)$ by the formula of Theorem 1.4. It would then be necessary to prove that $C(n, r)$ does indeed count the number of r-element subsets of a set of n elements. Our combinatorial approach is illustrated by our proofs of the next three theorems.

Theorem 1.5
For all $r, n \in \mathbb{N}$, with $r \leqslant n$, $C(n, r) = C(n, n - r)$.

Proof
Deciding which r objects to select from a set of n objects amounts to exactly the same thing as deciding which $n - r$ objects not to select. Hence the number of ways of choosing r objects from n is the same as the number of ways of choosing $n - r$ objects from n.

The next theorem helps explain how the binomial coefficients get their name.

Theorem 1.6 *(The Binomial Theorem)*
For all $n \in \mathbb{N}$,

$$(a + b)^n = a^n + C(n, 1)a^{n-1}b + C(n, 2)a^{n-2}b^2 + \ldots + b^n,$$

that is

$$(a + b)^n = \sum_{r=0}^{n} C(n, r)a^{n-r}b^r.$$

Proof
Consider the product

$$(a + b)(a + b) \dots (a + b)$$

with n sets of brackets. When we multiply out this product each separate term that arises comes from choosing either a or b from each bracket, and then multiplying these as and bs together. We obtain the term $a^{n-r}b^r$ if we choose b from r of these brackets and a from the remaining $n - r$ brackets. Thus the number of terms of the form $a^{n-r}b^r$ that we obtain equals the number of ways of choosing r brackets from which to pick b, and this number is $C(n, r)$. Hence, when we gather similar terms together, the coefficient of $a^{n-r}b^r$ is $C(n, r)$.

The idea we have used in giving this combinatorial proof of the Binomial Theorem will play an important role later in this book (in Chapter 5). The algebraic expression $(a + b)^n$ is called a **binomial** (from the late Latin *binomius* meaning 'having two personal names') and this is how the binomial coefficients get their name. The Binomial Theorem was first proved by Isaac Newton, though the bionomial coefficients were known and tabulated long before Newton's time. His main contribution was to prove the form of this theorem which applies when the exponent n is not a natural number.

The third combinatorial theorem about binomial coefficients leads to a very well-known method for calculating their values.

Theorem 1.7
For all $r, n \in \mathbb{N}$, with $r \leqslant n$,

$$C(n + 1, r) = C(n, r) + C(n, r - 1).$$

Proof
Again we emphasize that our aim is to give a combinatorial proof of this formula. An algebraic proof, using the formula for the binomial coefficients, is very straightforward, but hides the combinatorial meaning of the formula.

Let X be a set containing $n + 1$ objects. We count the number of ways of choosing a subset of r objects from X. Let a be one particular fixed element of X. We divide the r-element subsets of X into two classes.

The first class consists of those r-element subsets of X which do not include a. Such subsets are made up of r elements chosen from the n-element set $X \backslash \{a\}$, and hence there are $C(n, r)$ sets in this class.

The second class consists of those r-element subsets which include a. Such a subset consists of a and $r - 1$ elements chosen from the n-element set $X \backslash \{a\}$. Thus there are $C(n, r - 1)$ elements in this class.

These two classes between them include all the r-element subsets of X and no r-element subset is in both classes. Hence $C(n+1, r)$, the number of r-element subsets of X, is given by

$$C(n+1, r) = C(n, r) + C(n, r-1).$$

If the binomial coefficients are set out in rows, with the numbers $C(n, r)$, for $0 \leqslant r \leqslant n$, in the nth row, then it follows from Theorem 1.7 that each number in the $(n+1)$th row is the sum of the two adjacent numbers in the row above. Thus the binomial coefficients form an array, usually called **Pascal's Triangle**, after the seventeenth-century French mathematician Blaise Pascal, although the triangle was known much earlier, occurring, for example, in Chinese manuscripts. The first few rows of Pascal's Triangle are shown in Figure 1.1.

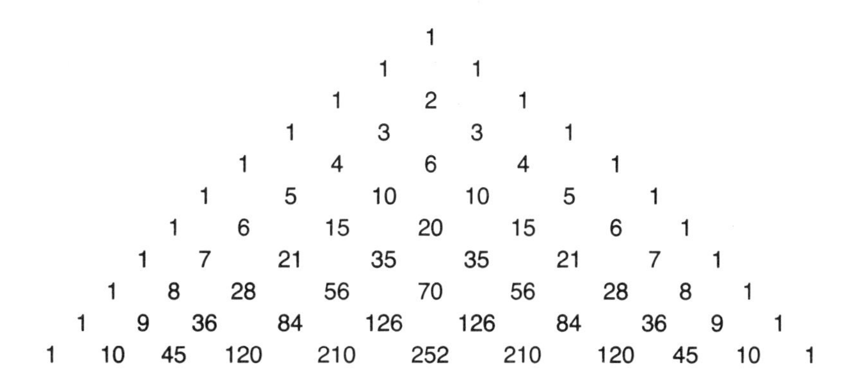

Fig. 1.1 Pascal's Triangle.

There are innumerable relationships between the binomial coefficients, a few of which can be found in the exercises at the end of this section. We conclude the section with a simple application of Theorem 1.4 to number theory.

Theorem 1.8
For each positive integer r, the product of r consecutive positive integers is divisible by $r!$.

Proof
We need to prove that for each positive integer k, the product $k(k+1)(k+2)\ldots(k+r-1)$ is divisible by $r!$, that is, that

$$\frac{k(k+1)(k+2)\ldots(k+r-1)}{r!}$$

is an integer. But this number is just the binomial coefficient, $C(k + r - 1, r)$, by Theorem 1.4. Since this binomial coefficient gives the number of r-element subsets of a $(k + r - 1)$-element set, it must be an integer.

Exercises

1.3.1. A mathematics course offers students the choice of three options from 12 courses in pure mathematics, two options from 10 courses in applied mathematics, two options from 6 courses in statistics, and one option from 4 courses in computing. In how many different ways can students choose their eight options?

1.3.2. If n points are placed on the circumference of a circle and all the lines connecting them are joined, what is the largest number of points of intersection of these lines inside the circle that can be obtained?

1.3.3. Prove that a set of n elements has altogether 2^n subsets, and deduce that for each positive integer n,

$$\sum_{r=0}^{n} C(n, r) = 2^n.$$

(You are encouraged to give combinatorial proofs, and not to use the Binomial Theorem.)

1.3.4. Prove that for each positive integer n,

$$\sum_{r=0}^{n} [C(n, r)^2] = C(2n, n).$$

1.3.5. Prove that for all $n, k, s \in \mathbb{N}$ with $s \leqslant k \leqslant n$,

$$C(n, k)C(k, s) = C(n, s)C(n - s, k - s).$$

(This is very easy to prove from the formula for the binomial coefficients, but you are again encouraged to find a combinatorial proof.)

1.3.6. Let X be a finite set. Prove that the number of subsets of X which contain an even number of elements is equal to the number of subsets of X which contain an odd number of elements. Deduce that for each positive integer n,

$$\sum_{r=0}^{n} (-1)^r C(n, r) = 0.$$

(Note that this last formula can be proved algebraically directly from the Binomial Theorem, by putting $a = 1$, and $b = -1$. The point of the question, however, is to give a combinatorial proof of this result.)

1.3.7. Suppose that we have k different sorts of objects, with an unlimited stock of each sort. Let $H(n, k)$ be the number of different ways of choosing n of these objects. Prove that for all $k, n \in \mathbb{N}$,

$$H(n, k) = C(n + k - 1, k).$$

(If we think of the k different sorts of objects as being k different algebraic symbols $x_1, \ldots x_k$, then choosing n of these objects corresponds to choosing a term of degree n built up from these symbols, that is, a term of the form

$$x_1^{n_1} x_2^{n_2} \ldots x_k^{n_k},$$

where $n_1, \ldots, n_k$ are natural numbers which add up to n. Thus $H(n, k)$ is the number of terms of degree n in k symbols.)

1.3.8. The Multinomial Theorem

Prove that the coefficient of $x_1^{k_1} x_2^{k_2} \ldots x_s^{k_s}$ in the expansion of

$$(x_1 + x_2 + \ldots + x_s)^n,$$

where $k_1 + k_2 + \ldots + k_s = n$, is

$$\frac{n!}{k_1! k_2! \ldots k_s!}$$

1.4 APPLICATIONS TO PROBABILITY PROBLEMS

We begin with a very simple problem, though it could lead us into deep water.

Problem 5

A fair coin is tossed ten times. What is the probability of getting four heads?

Solution

Each toss can result in one of two results, either a head or a tail. Thus for a sequence of ten tosses there are altogether

$$2^{10} = 1024$$

possible outcomes. The number of these sequences which consist of four heads and six tails is the same as the number of ways of choosing four of the ten tosses to be heads. Thus it is $C(10, 4) = 210$. Hence the probability (to three decimal places) of getting four heads is

$$\frac{210}{1024} = 0.205.$$

This calculation is straightforward enough, but what does it mean? And how did we know it should be done in this way?

Fortunately from the point of view of combinatorics we do not have to answer the difficult philosophical question as to what is meant by **probability**. Our work in this and similar probability problems begins after any philosophical work has been done, and a precise mathematical problem has been formulated. These problems will involve a set of events E, together with a subset E_1. The probability of an event falling into the subset E_1 is defined to be the ratio

$$\frac{\#(E_1)}{\#(E)}.$$

Our task will thus be to calculate the numbers that occur in this ratio, and we will not need to concern ourselves with its philosophical significance.

However there remains the question of how we knew, from the formulation of the problem, which the sets E and E_1 of events are in this particular case. You may have noticed that, in the statement of the problem, a 'fair coin' is referred to. This is intended to indicate that on any one throw of the coin the probability (whatever this means) of getting a head is exactly the same as the probability of getting a tail, and that the chance for any one toss is independent of what happens in earlier or later tosses. Thus any one sequence of outcomes is as likely as any other sequence. This probabilistic stipulation is reflected in our solution where we took the set of events, E, to be the set of all these possible sequences of outcomes.

For us, doing combinatorics, that is the end of the matter. If you wish to apply the result of this and similar calculations to practical situations, you need to know how realistic the assumption of a 'fair coin' is, and also how statements of probability are to be interpreted. These are not easy questions, and so it is fortunate that, in this book, I can largely avoid answering them.

In general the statement of a probability problem should indicate which set is to be taken as the set, E, of events. Thus the events in E are events which are to be regarded as equally probable. This indication is often done in a coded way. For example, many of these problems concern packs of cards 'dealt at random'. This is intended to mean that any of the 52! ways in which the pack can be ordered is as likely as any other. Hence the set E will consist of these 52! arrangements, or will be derived from it in some straightforward way.

The codes used in this way to set up probability problems are analogous to the coded way of describing problems in mechanics, where such phrases as 'a light string' or 'a frictionless pulley' are intended to indicate what assumptions can be made in devising the mathematical model of the situation.

The examples that follow in this section, and in the subsequent exercises, are mainly taken from card games, and, in particular, from the game of bridge. All you need to know about this game is that it is played with a standard pack of 52 cards. This pack is divided into four **suits**, spades (for which we use the symbol ♠), hearts (♡) diamonds (◇) and clubs (♣). There are 13 cards of each suit. Each card has a **rank**. The ranks of the 13 cards in each suit are 2, 3, 4, 5, 6, 7, 8, 9, 10, J (Jack), Q (Queen), K (King) and A (Ace).

In a game of bridge there are four players, often referred to as North, South, East and West. North and South play as partners, as do East and West. In the initial deal each player is dealt a hand of 13 cards. Thus the number of different bridge hands any one player can be dealt is the number of ways of choosing 13 cards from the full pack of 52 cards, which is

$$C(52, 13) = 635\,013\,559\,600.$$

A bridge hand will usually contain some cards from each of the four suits, but sometimes one, or more, of the suits will not be represented. When there are no cards of a particular suit we say that the hand has a **void suit**. For example, a hand with five spades, four diamonds and four clubs has a void heart suit. The problem we shall try to solve is to count the number of bridge hands which have at least one void suit. We begin with a rather easier problem.

Problem 6

How many bridge hands are there with five spades, four diamonds and four clubs?

Solutions

We can easily solve this problem by the methods of this chapter. To choose a hand of the kind described, we first choose five spades from the 13 spades in the pack, which can be done in $C(13, 5)$ ways. We then choose four diamonds, which we can do in $C(13, 4)$ ways, and then the four clubs, which also can be done in $C(13, 4)$ ways. Hence, using the Principle of Multiplication of Choices (see Theorem 1.1), the total number of hands with five spades, four diamonds and four clubs is

$$C(13, 5) \times C(13, 4) \times C(13, 4) = 657\,946\,575.$$

For future reference we give the values of $C(13,r)$ in Table 1.1.

A hand with five cards in one suit, four cards in two other suits, and hence no cards in the fourth suit, is said to have a 5-4-4-0 **suit distribution**. In general a hand with an *a-b-c-d* suit distribution, where

Table 1.1 Values of $C(13, r)$

r	0	1	2	3	4	5	6	7	8	9	10	11	12	13
$C(n, r)$	1	13	78	286	715	1287	1716	1716	1287	715	286	78	13	1

$a \geqslant b \geqslant c \geqslant d$, and $a + b + c + d = 13$, is a hand with a cards in one suit, b cards in another suit, c cards in a third suit and d cards in the fourth suit.

Problem 7
How many bridge hands are there with a 5-4-4-0 suit distribution?

Solution
We know from Problem 6 that once we have decided which suit is to contain five cards and which two suits are to contain four cards each, then there are 657 946 575 hands with these number of cards in those suits. So we need only count the number of ways of choosing the suits which have these particular numbers of cards in them. We can first choose the suit which is to have five cards in $C(4, 1)$ ways, as there are four suits to choose from. This leaves three suits from which to choose the two four-card suits which can be done in $C(3, 2)$ ways. These choices, of course, determine which suit is to have no cards in it. Hence there are $C(4, 1) \times C(3, 2) = 12$ ways of specifying which suits have the specified number of cards. It follows that there are

$$12 \times 657\,946\,575 = 7\,895\,358\,900$$

hands with a 5-4-4-0 suit distribution.

To work out the total number of bridge hands with at least one void suit, we need only work out those suit distributions which include one or more void suits. We can then use the method of Problem 7 to calculate the number of hands with each of these distributions. You are encouraged to at least try a few of these calculations for yourself, before turning to Table 1.2 where the number of hands with each of these suit distributions is given.

We see from Table 1.2 that there are 32 427 298 180 different bridge hands with at least one void suit. This is 5.1% of all the possible bridge hands. So if we assume that any one hand is as likely to be dealt as any other hand, the probability of having at least one void suit is 0.051 (to 3 decimal places).

Practical bridge players may wonder whether the theoretical assumption that any one deal is as likely to be dealt as any other is justified in practice. At the end of the play of one hand the cards are arranged in tricks of four cards, and bridge tricks tend to consist of cards all from

Table 1.2 Bridge hand suit distributions with at least one void suit

Suit distribution	*Number of hands*
13-0-0-0	4
12-1-0-0	2 028
11-2-0-0	73 008
11-1-1-0	158 184
10-3-0-0	981 552
9-4-0-0	6 134 700
10-2-1-0	6 960 096
8-5-0-0	19 876 428
7-6-0-0	35 335 872
9-2-2-0	52 200 720
9-3-1-0	63 800 880
8-4-1-0	287 103 960
6-6-1-0	459 366 336
8-3-2-0	689 049 504
7-5-1-0	689 049 504
7-3-3-0	1 684 343 232
7-4-2-0	2 296 831 680
6-5-2-0	4 134 297 024
5-5-3-0	5 684 658 408
5-4-4-0	7 895 358 900
6-4-3-0	8 421 716 160
Total	32 427 298 180

the same suit. So it might seem that unless the shuffling is done very carefully, the cards in the pack before the next deal are more likely to be next to a card of the same suit than in a random pack. This question is discussed in some detail in the standard work on the mathematics of bridge.* The conclusion drawn is that provided the pack is shuffled with reasonable care, the theoretical probabilities should be realized in practice.

The method we have used may seem rather long-winded (especially if you have checked all the calculations for yourself). Surely there must be a quicker way to do it. Here is an alternative approach. Let V_S, V_H, V_D and V_C be the sets of bridge hands with void spades, hearts, diamonds and clubs suits, respectively. As a hand in V_S contains no spades, it consists of 13 cards chosen from the remaining 39 cards in the pack. Thus there are $C(39, 13) = 8\,122\,454\,444$ hands in V_S. Likewise,

*Émile Borel and André Cheron, *Théorie mathématique du bridge*, 2nd edn, Paris, 1955.

V_H, V_D and V_C also contain this number of hands. Hence the total number of bridge hands with at least one void suit is

$$4 \times 8\,122\,425\,444 = 32\,489\,701\,776.$$

You will agree that this is a much quicker method. The only thing wrong with it is that it has not given the correct answer, although it is not far out. Where have we gone wrong?

It is not difficult to see where our mistake lies. Let V be the set of all bridge hands with at least one void suit. Then

$$V = V_S \cup V_H \cup V_D \cup V_C. \tag{1.1}$$

We assume that it follows from expression (1.1) that

$$\#(V) = \#(V_S) + \#(V_H) + \#(V_D) + \#(V_C). \tag{1.2}$$

(Recall that we are using $\#(X)$ for the number of elements in the set X.) In making the deduction from expression (1.1) to expression (1.2) we have overlooked the fact that some of the hands in V are in more than one of the sets V_S, V_H, V_D, V_C. For example, a hand with seven spades and six hearts is in both V_D and V_C. So in equation (1.2) we have counted some of the hands more than once, and it is no wonder that our answer turned out to be rather higher than the correct answer that we obtained by considering all the suit distributions separately.

The deduction from equation (1.1) and equation (1.2) would be valid if the sets concerned were pairwise disjoint. Thus if we have a collection of sets, $X_1, \ldots, X_k$, and if, for $1 \leqslant i < j \leqslant k$, $X_i \cap X_j = \varnothing$, then

$$\#\left(\bigcup_{i=1}^{k} X_i\right) = \sum_{i=1}^{k} \#(X_i). \tag{1.3}$$

Note that we have already tacitly used this formula, for example, in the proof of Theorem 1.7. In the next chapter we see how to modify equation (1.3) in the case where the sets are not disjoint. This will enable us to alter equation (1.2) so as to enable us to arrive at the correct answer.

Exercises

1.4.1. How many bridge hands are there with the following suit distributions?

(a) 5-4-3-1
(b) 4-4-3-2
(c) 4-4-4-1

1.4.2. In a bridge deal you and your partner between you have nine spades. What is the probability that the remaining four spades in your opponents' hands are divided equally between them? (There are very many more problems of a similar kind that can be asked about probabilities in the game of bridge. The book by Émile Borel and André Cheron, mentioned earlier, discusses many of these in detail.)

1.4.3. Poker hands

A poker hand consists of five cards drawn from the full pack of 52 cards. It can be of one of the following types:

(a) *Flush*: five cards all of the same suit, but not forming a sequence, e.g. 5♢, 7♢, J♢, Q♢, A♢.
(b) *Four of a kind*: Four cards of the same rank, plus one other card, e.g. J♠, J♡, J♢, J♣, 5♡.
(c) *Full house*: Three cards of one rank, and two cards of another rank, e.g. 7♠, 7♡, 7♣, 4♢, 4♣.
(d) *One pair*: Two cards of one rank, and three cards, all of other different ranks, e.g. 10♢, 10♣, Q♠, 6♡, 4♢.
(e) *Straight*: Five cards, not all of the same suit, whose ranks are consecutive (note that for this purpose an Ace may count as either high, as in the first example below, or low, as in the second example) e.g. 10♡, J♢, Q♢, K♣, A♠ or A♡, 2♣, 3♢, 4♣, 5♡.
(f) *Straight flush*: Five cards of the same suit whose ranks are consecutive (again, an Ace may count either high or low) e.g. 4♣, 5♣, 6♣, 7♣, 8♣.
(g) *Three of a kind*: Three cards of one rank and two cards of other different ranks, e.g. 5♠, 5♡, 5♢, 9♠, J♣.
(h) *Two pairs*: Two cards of one rank, two cards of another rank, and a fifth card of another rank, e.g. J♡, J♢, 7♠, 7♣, 5♡.
(i) *Other hands*: All the other hands which do not come into any of the above categories.

Calculate how many poker hands there are of each of the above kinds. Hence work out for each kind, the probability that a poker hand dealt at random is of that kind.

1.4.4. The different types of poker hands are ranked according to their probabilities, so that, for example three of a kind beats one pair. After the hands have been dealt the players have the chance to exchange one or more cards to try to improve their hands. Suppose that you hold the following hand:

J♠, J♣, 8♢, 5♡, 4♣.

Are you more likely to improve your hand by keeping just the two Jacks, and exchanging three cards, or by keeping J♠, J♣ and 8♢ and exchanging just two cards?

1.4.5. A bag contains 50 red balls and 50 blue balls. Ten balls are drawn at random from the bag. What is the probability that this sample will contain five red balls and five blue balls? (This problem is connected with reliability of opinion polls. See the solution for more about this.)

1.4.6. If there are n people in a room, what is the probability that at least two of them share a birthday? How large does n have to be before this probability becomes at least one half? (By 'sharing a birthday' we mean that two people were born on the same day of the same month, but not necessarily in the same year. For the purpose of this problem ignore leap years, so that there are 365 possible birthdays for each person. You should also assume that all these birthdays are equally likely (in fact, as Exercise 1.4.7 shows, if some dates are more likely to be birthdays than others, then the probability of a coincidence increases.) This amounts to taking the set, E, of events to be the set of all possible combinations of birthdays for the set of n people, and taking E_1 to be the subset where at least two of these birthdays are the same.)

1.4.7. Suppose there are $2n$ balls in a bag, of which a are red and b are blue, with $a + b = 2n$. One ball is removed, at random, from the bag, and then replaced. Then another ball is drawn, at random, from the bag. Calculate the probability that the two balls you have sampled are of the same colour. Show that this probability is a minimum when $a = b = n$.

1.4.8. How many people do there need to be in a room before there is a probability greater than 0.5 that at least three of them share a birthday?

2

The inclusion–exclusion principle

2.1 DOUBLE COUNTING

We saw at the end of chapter 1 that the formula

$$\#(A \cup B \cup C \cup \ldots) = \#(A) + \#(B) + \#(C) + \ldots$$

holds provided that the sets A, B, $C, \ldots$ are disjoint, but not if there is any overlap among them. In this section we develop the formula which holds in the case of a general union of sets which are not necessarily disjoint. We begin with the easiest situation to understand, where we have just two sets, say A and B.

In the sum $\#(A) + \#(B)$ we count each element of A once and each element of B once. So the elements of $A \cap B$ get counted twice, once through their membership of A and once through their membership of

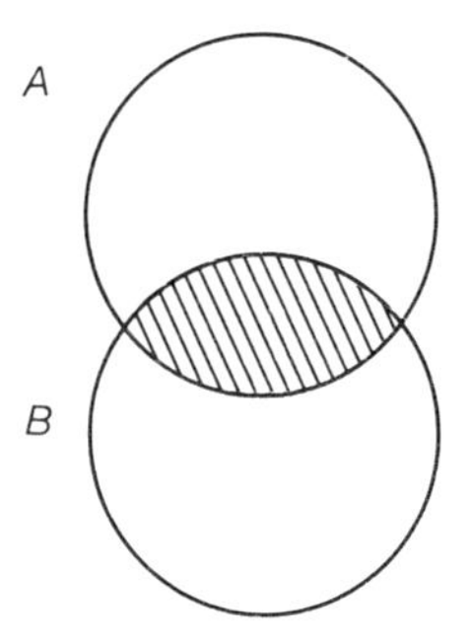

B. To obtain the number of elements in $A \cup B$ we therefore have to subtract the number of elements which are counted twice, namely, those elements which are both in A and in B. Thus we obtain

Theorem 2.1
For all sets A and B,

$$\#(A \cup B) = \#(A) + \#(B) - \#(A \cap B).$$

Problem 1

How many integers are there in the range from 1 to 1 000 000 which are divisible by 2 or by 3 or both?

Solution

We let D_2 and D_3 be the sets of those integers in the range from 1 to 1 000 000 which are divisible by 2 and by 3, respectively. By Theorem 2.1,

$$\#(D_2 \cup D_3) = \#(D_2) + \#(D_3) - \#(D_2 \cap D_3). \qquad (2.1)$$

Clearly, $\#(D_2) = 500\,000$ and $\#(D_3) = 333\,333$. An integer is divisible by both 2 and 3 if and only if it is divisible by 6. Hence $\#(D_2 \cap D_3) = 166\,666$. Thus, by equation (2.1),

$$\#(D_2 \cup D_3) = 500\,000 + 333\,333 - 166\,666 = 666\,667.$$

Theorem 2.1 can easily be extended to deal with the case of three sets. All we need do is to write $A \cup B \cup C$ as $(A \cup B) \cup C$ and then apply Theorem 2.1 to this union of two sets, making use of some elementary set algebra. In this way we obtain

$$\begin{aligned}\#(A \cup B \cup C) &= \#((A \cup B) \cup C)\\ &= \#(A \cup B) + \#(C) - \#((A \cup B) \cap C),\\ &= \#(A) + \#(B) - \#(A \cap B) + \#(C)\\ &\quad - \#(A \cup B) \cap C)),\end{aligned}$$

using Theorem 2.1 twice. Now, by the distributive law for unions and intersections of sets,

$$(A \cup B) \cap C = (A \cap C) \cup (B \cap C).$$

Hence, using Theorem 2.1 again,

$$\begin{aligned}\#((A \cup B) \cap C &= \#(A \cap C) + \#(B \cap C)\\ &\quad - \#((A \cap C) \cap (B \cap C))\\ &= \#(A \cap C) + \#(B \cap C) - \#(A \cap B \cap C).\end{aligned}$$

We can therefore deduce that

$$\begin{aligned}\#(A \cup B \cup C) &= (\#(A) + \#(B) + \#(C))\\ &\quad - (\#(A \cap B) + \#(A \cap C) + \#(B \cap C)) + \#(A \cap B \cap C).\end{aligned}$$

Now let us see what this formula means. The three terms in the first bracket count the elements of A, B and C separately. Thus when we add them up, elements which are in two of these sets are counted twice, and those in all three are counted three times. We take account of the double counting by subtracting the terms in the second bracket.

However, when we do this, an element which is in all three of the sets is discounted three times. To compensate for this we add on the final term $\#(A \cap B \cap C)$. We can illustrate this process by the following diagrams:

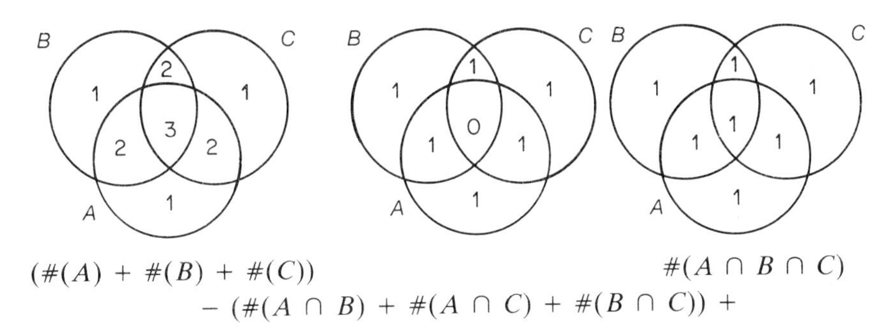

$(\#(A) + \#(B) + \#(C))$ $\quad - (\#(A \cap B) + \#(A \cap C) + \#(B \cap C)) +$ $\quad \#(A \cap B \cap C)$

We see from these diagrams that our formula does indeed count each element of $A \cup B \cup C$ exactly once. Notice that this is accomplished by alternating including and excluding elements from the count.

It is now not difficult to see how to extend this formula to deal with the general case of the number of elements in a union of n sets. The only complication is with the notation needed to deal with the most general situation. The theorem gets its name from the inclusion and exclusion process corresponding to the alternate + and − signs.

Theorem 2.2 *(The inclusion–exclusion theorem)*
For all sets $A_1, A_2, \ldots, A_n$,

$$\begin{aligned}\#(A_1 \cup A_2 \cup \ldots \cup A_n) = {} & \big(\#(A_1) + \#(A_2) + \ldots + \#(A_n)\big) \\ & - \big(\#(A_1 \cap A_2) + \#(A_1 \cap A_3) + \ldots + \#(A_{n-1}) \cap (A_n)\big) \\ & + \big((\#(A_1 \cap A_2 \cap A_3) + \#(A_1 \cap A_2 \cap A_4) \\ & + \ldots . + \#(A_{n-2} \cap A_{n-1} \cap A_n)\big) \\ & \vdots \\ & + (-1)^{n+1}\, \#(A_1 \cap A_2 \cap \ldots A_n). \end{aligned} \tag{2.2}$$

A word about the notation is necessary before we indicate the proof of this theorem. It is intended to indicate that the second bracket contains terms corresponding to each pair of sets, the third bracket terms corresponding to each triple of sets, and so on. If we want to be more explicit about this, at the price of complicating our formula, we can write equation (2.2) as follows:

$$\#\left(\bigcup_{i=1}^{n} A_i\right) = \sum_{k=1}^{n} (-1)^{k+1} \left(\sum_{1 \leq i_1 < \ldots < i_k \leq n} \#(A_{i_1} \cap \ldots \cap A_{i_k})\right) \tag{2.3}$$

Here the expression $1 \leqslant i_1 < \ldots < i_k \leqslant n$ under the summation symbol is intended to indicate that the sum is taken over all choices of integers $i_1, \ldots, i_k$ which satisfy these inequalities. Thus it is a way of saying that we sum over all the k-element subsets, $\{i_1, \ldots, i_k\}$ of the set $\{1, 2, \ldots, n\}$.

We are now ready for the proof.

Proof

Take an element $x \in A_1 \cup \ldots \cup A_n$. We calculate how many times x is included and excluded by the formula on the right-hand side of equation (2.3).

Suppose that x occurs in exactly m of the sets A_i, say $A_{j_1}, \ldots, A_{j_m}$. Then $x \in A_{i_1} \cap \ldots \cap A_{i_k}$ if and only if

$$\{i_1, \ldots, i_k\} \subseteq \{j_1, \ldots, j_m\}. \tag{2.4}$$

This can occur only if $1 \leqslant k \leqslant m$, in which case there are $C(m, k)$ ways of choosing $i_1, \ldots, i_k$ so that (2.4) holds. Hence x is counted $C(m, k)$ times in the sum

$$\sum_{1 \leqslant i_1 < .. < i_k \leqslant n} \#(A_{i_1} \cap \ldots \cap A_{i_k})$$

which occurs in the bracket on the right-hand side of equation (2.3). Thus, taking account of the alternating signs, x is counted

$$\sum_{k=1}^{m} (-1)^{k+1} C(m, k) \tag{2.5}$$

times by the formula on the right-hand side of equation (2.3). Now,

$$\sum_{k=0}^{m} (-1)^k C(m, k) = 0$$

(see Exercises 1.3.6), from which it follows that

$$C(m, 0) = \sum_{k=1}^{m} (-1)^{k+1} C(m, k).$$

and hence that expression (2.5) equals 1. Hence each element is counted just once in formula (2.3), and so this formula does indeed give the number of elements in $A_1 \cup \ldots \cup A_n$.

We can now use the Inclusion–Exclusion Theorem to modify the calculation of the number of bridge hands with at least one void suit at the end of Chapter 1. Remember that V_S, V_H, V_D and V_C are the sets of hands with no spades, hearts, diamonds and clubs, respectively. By the Inclusion–Exclusion Theorem,

$$\begin{aligned}\#(V_S \cup V_H \cup V_D \cup V_C) = {} & (\#(V_S) + \#(V_H) + \#(V_D) + \#(V_C)) \\ & - (\#(V_S \cap V_H) + \ldots + \#(V_D \cap V_C)) \\ & + (\#(V_S \cap V_H \cap V_D) + \ldots + \\ & \#(V_H \cap V_D \cap V_C)) \\ & - \#(V_S \cap V_H \cap V_D \cap V_C). \qquad (2.6)\end{aligned}$$

A hand in V_S contains 13 cards drawn from the 39 cards in the pack which are not spades. Hence $\#(V_S) = C(39, 13)$, and similarly for each of the terms in the first bracket in equation (2.6). The terms in the second bracket are also all equal. For example, $V_S \cap V_H$, consists of hands of 13 cards drawn from the 26 cards in the pack which are neither spades nor hearts. So each term in the second bracket equals $C(26, 13)$. Likewise, all the terms in the third bracket are equal to $C(13, 13)$. There are, of course, no hands containing a void in all four suits, so $\#(V_S \cap V_H \cap V_D \cap V_C) = 0$. Thus it follows from equation (2.6) that

$$\begin{aligned}\#(V_S \cup V_H \cup V_D \cup V_C) &= 4 \times C(39, 13) - 6 \times C(26, 13) \\ &\quad + 4 \times C(13, 13) \\ &= 32\,427\,298\,180.\end{aligned}$$

This does agree with our first, more long-winded calculation in chapter 1.

The next problem involves an application of the Inclusion–Exclusion Theorem of a rather different kind.

Problem 2

A set of n objects is sampled at random with replacement. What is the probability that after s samples have been drawn each object has been sampled at least once?

Solution

'Sampling with replacement' means that the set from which the samples are drawn remains the same throughout. It occurs in many different contexts. For example, both bird watchers and train spotters can be thought of as sampling with replacement. Also, when you throw dice, you are sampling the numbers on the dice with replacement, since the set of numbers that can come up on each throw stays the same.

The statement that the objects are sampled *at random* can be interpreted as meaning that any sequence of s samples is as likely as any other. So the required probability is just the ratio

$$\theta_n(s) = \frac{\#(\text{Sequences of } s \text{ samples in which each object occurs})}{\#(\text{All sequences of } s \text{ samples})}. \qquad (2.7)$$

The number in the denominator of equation (2.7) is very easy to calculate. Each time we draw a sample we have n objects to choose from. Hence a sequence of s samples can be selected in n^s ways.

To calculate the number in the numerator, it is easier first to calculate the number of sequences of s samples in which at least one of the objects does not occur. We let A_i be the set of those sequences of s samples in which the ith object is missing. The number in the numerator of equation (2.7) is thus

$$n^s - \#\left(\bigcup_{i=1}^{n} A_i\right),$$

and by the Inclusion–Exclusion Theorem,

$$\#\left(\bigcup_{i=1}^{n} A_i\right) = \sum_{k=1}^{n} (-1)^{k+1} \left(\sum_{1 \leq i_1 .. < i_k \leq n} \#(A_{i_1} \cap \ldots \cap A_{i_k})\right) \quad (2.8)$$

$(A_{i_1} \cap \ldots \cap A_{i_k})$ is the set of those sequences of s samples in which k of the objects do not occur, and hence in which each sample can be chosen in $(n - k)$ ways. Hence

$$\#(A_{i_1} \cap \ldots \cap A_{i_k}) = (n - k)^s.$$

There are $C(n, k)$ terms in the kth bracket of equation (2.8) and hence

$$\#\left(\bigcup_{i=1}^{n} A_i\right) = \sum_{k=1}^{n} (-1)^{k+1} C(n, k)(n - k)^s.$$

Thus the number in the numerator of equation (2.7) is

$$n^s - \sum_{k=1}^{n} (-1)^{k+1} C(n, k)(n - k)^s = \sum_{k=0}^{n} (-1)^k C(n, k)(n - k)^s,$$

and so the required probability, $\theta_n(s)$, is given by

$$\theta_n(s) = \frac{1}{n^s} \sum_{k=0}^{n} (-1)^k C(n, k)(n - k)^s. \quad (2.9)$$

There is no straightforward way to simplify this formula in general. However in two special cases we can do this.

Corollary 2.3
For $s < n$,

$$\sum_{k=0}^{n} (-1)^k C(n, k)(n - k)^s = 0.$$

Proof
When $s < n$ there cannot be a sequence of s samples which includes all the n objects. So in this case $\theta_n(s) = 0$.

Corollary 2.4

$$\sum_{k=0}^{n}(-1)^k C(n, k)(n - k)^n = n!$$

Proof

When $s = n$, a sequence of s samples containing each of the n objects, is just a permutation of those objects, and we know that there are $n!$ permutations of n objects.

Exercises

2.1.1. How many integers are there in the range from 1 to 1 000 000 which are either perfect squares or perfect cubes, or both?

2.1.2. How many integers are there in the range from 1 to 1 000 000 which are divisible by none of 2, 3, 5 and 7?

2.1.3. Euler's ϕ-function is defined by

$$\phi(n) = \#(\{k \in \mathbb{N}^+ : k \leqslant n \text{ and } k \text{ has no prime factors in common with } n\})$$

For example, $\phi(12) = 4$, since the positive integers up to and including 12 which have no prime factors in common with 12 are 1, 5, 7 and 11.

Use the Inclusion–Exclusion Theorem to find a formula for $\phi(n)$ when the distinct prime factors of n are $p_1, \ldots, p_k$, and show that

$$\phi(n) = n\left(1 - \frac{1}{p_1}\right)\left(1 - \frac{1}{p_2}\right)\cdots\left(1 - \frac{1}{p_k}\right).$$

2.1.4. How many times do you need to throw a dice so that there is a probability greater than 0.5 that each of the numbers 1, 2, 3, 4, 5, 6 is thrown at least once?

2.2 DERANGEMENTS

We begin with another problem involving a pack of cards.

Problem 3

Two packs of 52 cards are dealt simultaneously, one card at a time from each pack. What is the probability that at least once the same card is dealt from both packs?

Solution

Before reading this solution try to estimate the probability or, better still, solve the problem for yourself.

We can calculate the required probability, which we will denote p_{52}, by assuming that the order of the cards in one of the packs is fixed, and then counting the number of arrangements of the second pack which leads to at least one coincidence. The required probability will then be this number, divided by the total number of arrangements of the cards in the pack, which is, of course, 52!

We let A_i be the set of those arrangements of the second pack in which the ith card coincides with the ith card of the first pack. The number we want to calculate is thus

$$\#\left(\bigcup_{i=1}^{52} A_i\right).$$

so that

$$p_{52} = \frac{\#(\bigcup_{i=1}^{52} A_i)}{52!} \tag{2.10}$$

By the Inclusion–Exclusion Theorem

$$\#\left(\bigcup_{i=1}^{52} A_i\right) = \sum_{k=1}^{52} (-1)^{k+1}\left(\sum_{1 \leqslant i_1 < .. < i_k \leqslant 52} \#(A_{i_1} \cap \ldots \cap A_{i_k})\right). \tag{2.11}$$

Now $A_{i_1} \cap \ldots \cap A_{i_k}$ is the set of those arrangements of the second pack in which the cards in positions $i_1, \ldots, i_k$ are determined, as they must be the same as the corresponding cards in the first pack. The other $(52 - k)$ cards can be arranged in any order in the remaining $(52 - k)$ positions. Thus

$$\#(A_{i_1} \cap \ldots \cap A_{i_k}) = (52 - k)!.$$

The number of terms in the kth bracket of equation (2.11) is $C(52, k)$ and it therefore follows that

$$\#\left(\bigcup_{i=1}^{52} A_i\right) = \sum_{k=1}^{52} (-1)^{k+1} C(52, k)(52 - k)!$$

$$= \sum_{k=1}^{52} (-1)^{k+1} \frac{52!}{k!}. \tag{2.12}$$

Thus the required probability, which is given by substituting from equation (2.12) into equation (2.10), is

$$p_{52} = \sum_{k=1}^{52} (-1)^{k+1} \frac{1}{k!}. \tag{2.13}$$

For a pack of n cards all we would need to do is substitute n for 52 throughout the solution to this problem.

We can easily relate the number given by equation (2.13) to the number e. Using the series for e^x in the case where $x = -1$, we have

$$e^{-1} = \sum_{k=0}^{\infty} \frac{(-1)^k}{k!},$$

and hence

$$1 - e^{-1} = \sum_{k=1}^{\infty} \frac{(-1)^{k+1}}{k!}. \tag{2.14}$$

Comparing equations (2.13) and (2.14), we see that the former consists of the first 52 terms of the infinite series in the latter. This series converges so rapidly that $1 - e^{-1}$ provides a very accurate approximation to the required probability.

Again, if we were dealing with a pack of n cards the only change we would need to make in our answer would be to substitute n for 52 in equation (2.13). So the probability is given by the first n terms of the series for $1 - e^{-1}$. The rapid convergence of this series means that once we have at least six cards in the pack, the probability of at least one coincidence does not change appreciably as the number of cards increases. We denote this probability, for an n-card pack, by p_n. Some of values of p_n are given in Table 2.1, with the value of $1 - e^{-1}$ for comparison at the end (all to nine decimal places). The rapidity of the convergence of p_n to $1 - e^{-1}$ is evident from the values in this table.

Table 2.1 The probability, p_n, of dealing the same card at least once from two packs of n cards dealt simultaneously. As $n \to \infty$, the probability tends to $1 - e^{-1}$.

n	p_n
1	1.000000000
2	0.500000000
3	0.666666667
4	0.625000000
5	0.633333333
6	0.631944444
7	0.632142857
8	0.632118056
9	0.632120811
10	0.632120536
11	0.632120561
12	0.632120559
13	0.632120559
$1 - e^{-1}$	0.632120559

An arrangement of n objects in which no object occupies its original position is called a **derangement** of those objects. Problem 3 involves counting the number of arrangements of the second pack which are *not* derangements of the first pack. Using the method of that problem we can easily arrive at the following result, whose proof is left as an exercise.

Theorem 2.5
The number of derangements of a set of n objects is given by

$$n! \sum_{k=0}^{n} \frac{(-1)^k}{k!}.$$

The series in this expression consists of the first $n + 1$ terms of the infinite series for e^{-1}, hence the number given by the expression is approximately $n!e^{-1}$, and this approximation is very accurate for $n \geqslant 6$.

Exercises

2.2.1. If ten letters are placed in ten addressed envelopes at random, what is the probability that every letter is put into the wrong envelope?

2.2.2. Give the proof of Theorem 2.5.

2.2.3. For the purpose of this question by an **anagram** of a given word we will mean a rearrangement of the letters so that in each position there is a change of letter. Thus when all the letters of a word are different, an anagram is just the same as a derangement of its letters, but when some letters are repeated, not all the derangements are anagrams. For example, the word NOON has just the one anagram ONNO and the word GOOD has the two anagrams ODGO and OGDO, whereas SEE has no anagrams at all, as in any rearrangement of its letters, there is bound to be an E remaining in one of the two original positions where an E occurs.
How many anagrams do each of the following words have?
(i) RADAR
(ii) ANAGRAM.

3

Partitions

3.1 WHAT ARE PARTITIONS?

In this chapter we introduce the idea of a partition of a number. We explain what a partition is in the next paragraph. It is a fairly natural idea, interesting both in itself and because of its connection with other problems. So, one way or another, partitions will be involved in most of what we do in the rest of this book. In fact, we have already met some partitions, in the guise of the suit distributions that we discussed in Chapter 1. Here is the general definition.

A **partition** of the number n, where $n \in \mathbb{N}^+$, is a representation of n as a sum of positive integers. For example, the partitions of the number 5 are as follows:

$$\begin{gathered} 5 \\ 4+1 \\ 3+2 \\ 3+1+1 \\ 2+2+1 \\ 2+1+1+1 \\ 1+1+1+1+1 \end{gathered}$$

making seven of these partitions in all. Note that the order in which the numbers occur in the sum is not important. Two partitions count as being the same if they are made up of the same numbers in different orders. Thus, we regard $2+2+1$ and $2+1+2$ as being different ways of writing the same partition of 5. We will adopt the convention of writing the numbers in the partition in decreasing order. So the partition $2+1+2$ will usually be written as $2+2+1$.

The numbers which occur in a partition are called the **parts** of the partition. We use $p(n)$ to stand for the number of different partitions of n. The numbers $p(n)$ are sometimes referred to as the **unrestricted partition numbers**, marking the fact that $p(n)$ counts the total number of partitions of n, without any restriction except for the understanding that we mean partitions into positive integers. We are going to spend a good deal of time studying these unrestricted partition numbers.

Many different restrictions can be placed on partitions. We shall meet restrictions of the following kinds.

(a) Restrictions on the number of parts that can occur in a partition.
(b) Restrictions on the size of the parts (i.e. the magnitude of the numbers) that can occur in a partition.
(c) Restrictions on the number of repetitions that can occur in a partition.
(d) Restrictions on the type of numbers that can occur in a partition.

For example, a restriction of category (d) could be that all the numbers occurring in the partition are odd numbers, or square numbers or prime numbers. We obtain different sequences of partition numbers according to which restrictions are imposed. It is possible to impose more than one restriction at the same time.

Here we are getting close to the ill-defined boundary between combinatorics and number theory. For example, if we impose both the restriction that there are at most two parts and that these parts must be prime numbers, then the question whether $p_G(2n) \geqslant 1$, for all $n \in \mathbb{N}$ with $n > 1$, where p_G counts the number of partitions of this type, is just the question whether every even number greater than 2 is the sum of two prime numbers, which is Goldbach's well-known problem, as yet unsolved.

Coming back down to earth, notice that the bridge hand suit distributions that we discussed in section 1.4 correspond to partitions of the number 13 where we add the restriction that there are at most four positive integers making up the partitions. We use the notation $p_k(n)$ for the number of partitions of n into at most k parts. Although we have defined partitions to be partitions of *positive integers*, we will see that it is convenient to adopt the conventions that, for all $k \in \mathbb{N}^+$,

$$p(0) = 1 \text{ and}$$

$$p_k(0) = 1.$$

We can see from our list of partitions of the number 5 at the beginning of this chapter that

$$p_1(5) = 1$$

$$p_2(5) = 3$$

$$p_3(5) = 5$$

$$p_4(5) = 6$$

$$p_5(5) = 7.$$

Also, it can be seen from Table 1.2 that

$$p_3(13) = 21,$$

there being 21 suit distributions with at least one void suit.

The following lemma gives some elementary facts about these restricted partition numbers.

Lemma 3.1

For each $n \in \mathbb{N}^+$,

(a) $$p_1(n) = 1.$$

(b) $$\text{For all } k \geqslant n,\ p_k(n) = p(n).$$

(c) $$p_2(n) = \operatorname{int}\left(\frac{n}{2} + 1\right).$$

(Recall that $\operatorname{int}(x)$ is the integer part of x.)

Proof

(a) is obvious. (b) holds because there can be at most n parts in a partition of n.

We prove (c) by considering two separate cases. First, suppose that n is even, say $n = 2m$, with $m \in \mathbb{N}^+$. The distinct partitions of n into at most two parts are, in this case

$$\begin{gathered} 2m \\ (2m - 1) + 1 \\ (2m - 2) + 2 \\ \vdots \\ m + m \end{gathered}$$

making $m + 1$ different partitions. Hence

$$p_2(n) = m + 1 = \operatorname{int}\left(\frac{n}{2} + 1\right).$$

Second, suppose that n is odd, say $n = 2m + 1$. In this case the partitions of n into at most two parts are

$$\begin{gathered} 2m + 1 \\ (2m) + 1 \\ (2m - 1) + 2 \\ \vdots \\ (m + 1) + m \end{gathered}$$

and hence $p_2(n) = m + 1 = \operatorname{int}(n/2 + 1)$.

Exercises

3.1.1. Calculate the values of $p(n)$ for $n = 1, 2, 3, \ldots, 8$.

3.1.2. Calculate the values of $p_3(n)$ for $n = 1, 2, 3, \ldots, 10$.
3.1.3. Show that the number of **ordered partitions** of n, that is, partitions of n into positive integers when partitions which are made up of the same numbers, *but in a different order*, are counted as different (e.g., $3+2+1$ and $2+1+3$ count as different partitions of 6) is 2^{n-1}.

3.2 DOT DIAGRAMS

As well as restricting partitions according to the *number* of parts making them up, we can also put a restriction on the *size* of the parts, that is, on the positive integers which occur in the partition. We use $q_k(n)$ to stand for the number of partitions of n into parts of size at most k. Looking again at our list of partitions of the number 5 (section 3.1), we obtain the following values.

$$q_1(5) = 1$$
$$q_2(5) = 3$$
$$q_3(5) = 5$$
$$q_4(5) = 6$$
$$q_5(5) = 7.$$

It should strike you that these $q_k(n)$ values are the same as the corresponding $p_k(n)$ values that we found in section 3.1. This is not a coincidence. The proof of this uses the idea of a **dot diagram**; a very simple idea, but one which is surprisingly useful.

We can represent a partition by rows of dots corresponding to the parts making up the partitions. For example the partition

$$7 + 4 + 3 + 3 + 1$$

of 18 can be represented by the following dot diagram, where the number of dots in the row correspond to the terms of the partition:

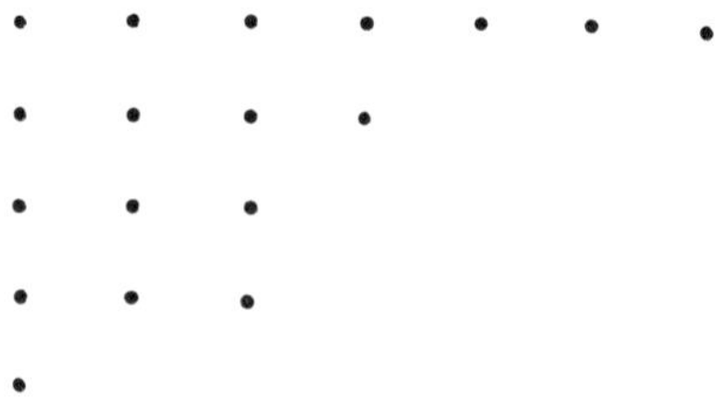

```
•  •  •  •  •  •  •
•  •  •  •
•  •  •
•  •  •
•
```

If we read this diagram vertically instead of horizontally, we obtain another partition:

$$5 + 4 + 4 + 2 + 1 + 1 + 1.$$

Since the number of dots has not changed this is also a partition of 18. We call this second partition the **dual** of the first partition. It is not difficult to see how a partition is related to its dual.

Lemma 3.2
The dual of the partition

$$k_1 + k_2 + \ldots + k_s \text{ with } k_1 \geqslant k_2 \geqslant \ldots \geqslant k_s,$$

is the partition

$$l_1 + l_2 + \ldots + l_t,$$

where $t = k_1$ and for, $1 \leqslant i \leqslant t$,

$$l_i = \#(\{j : 1 \leqslant j \leqslant s \text{ and } k_j \geqslant i\}). \tag{3.1}$$

Proof
The number of parts in the dual partition is equal to the number of columns in the dot diagram, which is the number of dots in the longest row of the dot diagram, thus $t = k_1$.

The number l_i is the number of dots in the ith column from the left. This is equal to the number of rows containing at least i dots, and this is given by formula (3.1) above.

Note that, in particular, it follows that $l_1 = s$.

We see from this that the number of parts in the first partition equals the size of the largest part in the dual partition, and vice versa. Thus, for each positive integer k, there is a one-to-one correspondence between the partitions of n into at most k parts and the partitions of n into parts of size at most k. In other words, we have proved the following result.

Theorem 3.3
For all $k, n \in \mathbb{N}^+$,

$$p_k(n) = q_k(n).$$

One advantage of this theorem is that arguments involving the size of the largest part are often easier to handle than those involving the number of parts. Thus it is often easier to prove results about $q_k(n)$ than about $p_k(n)$, even though, by the theorem, these numbers are equal. An example of this is given by the next theorem.

Theorem 3.4
For all $k, n \in \mathbb{N}$, with $k \leqslant n$,

$$\text{(a) } q_k(n) = q_{k-1}(n) + q_k(n - k),$$

and

$$\text{(b)}\ p_k(n) = p_{k-1}(n) + p_k(n-k).$$

Proof

We prove (a), and then (b) follows immediately, by Theorem 3.3.

We split the partitions of n into parts of size at most k into two classes. The first class consists of those partitions of this type which do not actually contain any part of size k. Thus this class is made up of partitions of n into parts of size at most $k-1$. It therefore contains $q_{k-1}(n)$ partitions.

The second class consists of those partitions of n into parts of size at most k which do contain at least one part of this size. When this part of size k is removed what remains is a partition of $n-k$ into parts of size at most k. Conversely, given a partition of $n-k$ into parts of size at most k, by adding another part of size k, we obtain a partition of n into parts of size at most k and containing at least one part of this size. Thus the partitions in this second class are in one-to-one correspondence with the partitions of $n-k$ into parts of size at most k, and thus this class contains $q_k(n-k)$ partitions.

These two classes are disjoint and cover all the partitions of n into parts of size at most k. Equations (a) therefore follows. Then equation (b) follows immediately from Theorem 3.3.

It should be noted that equations (a) and (b) of this theorem only hold in the case that $k = n$ because of our convention that $p_k(0) = q_k(0) = 1$.

The usefulness of Theorem 3.4 will become apparent later in this chapter. It can be used to derive some further relationships involving the partition numbers $q_k(n)$ and $p_k(n)$.

Theorem 3.5

For all $k, n \in \mathbb{N}$ with $k \leqslant n$,

$$\text{(a)} \qquad q_k(n) = q_{k-1}(n) + q_{k-1}(n-k) + q_{k-1}(n-2k) + \ldots + q_{k-1}(n-sk)$$

and

$$\text{(b)} \qquad p_k(n) = p_{k-1}(n) + p_{k-1}(n-k) + p_{k-1}(n-2k) + \ldots + p_{k-1}(n-sk)$$

where $s = \operatorname{int}(n/k)$.

Proof

By Theorem 3.4,

$$q_k(n) = q_{k-1}(n) + q_k(n-k) \tag{3.2}$$

Applying this theorem again (provided that $2k \leqslant n$),

$$q_k(n - k) = q_{k-1}(n - k) + q_k(n - 2k) \tag{3.3}$$

By equations (3.2) and (3.3)

$$q_k(n) = q_{k-1}(n) + q_{k-1}(n - k) + q_k(n - 2k).$$

Continuing in this way, we obtain

$$q_k(n) = q_{k-1}(n) + q_{k-1}(n - k) + q_{k-1}(n - 2k) + \ldots + q_{k-1}(n - (s - 1)k) + q_k(n - sk) \tag{3.4}$$

Since $s = \operatorname{int}(n/k)$, $n - sk$ is the remainder when n is divided by k, and thus $n - sk < k$. Therefore there are no partitions of $n - sk$ containing parts of size k (or larger). It follows that

$$q_k(n - sk) = q_{k-1}(n - sk). \tag{3.5}$$

Part (a) now follows at once from equations (3.4) and (3.5), and equation (b) by Theorem 3.3.

A natural question to ask at this stage is whether there is a formula for the values of $p(n)$, or $p_k(n)$. The answer will depend on what we mean by a formula, and to this second question we now turn.

Exercises

3.2.1. We have seen that the values of the function p_2 are given by

$$p_2(n) = \operatorname{int}\left(\frac{n}{2} + 1\right).$$

Show that, similarly, there is a quadratic $an^2 + bn + c$ such that for all $n \in \mathbb{N}$,

$$p_3(n) = \operatorname{int}(an^2 + bn + c).$$

(It is not difficult to find the quadratic empirically, by trying to fit the values of p_3, which are easy to calculate. To prove that the formula is correct, it is helpful to use the formula given in Theorem 3.4.)

3.2.2. Let $s_k(n)$ be the number of partitions of n whose smallest part is of size k. Thus $s_k(n)$ counts the partitions of n in which the smallest number that occurs is k. For example, $s_2(9) = 4$ because the partitions of 9 in which the smallest part is of size 2 are $7 + 2$, $5 + 2 + 2$, $4 + 3 + 2$ and $3 + 2 + 2 + 2$.

(a) Calculate the values of $s_k(10)$ for $1 \leqslant k \leqslant 8$.

(b) Prove that for all $n \in \mathbb{N}^+$,

$$p(n) = 1 + \sum_{k=1}^{\operatorname{int}(n/2)} s_k(n).$$

(c) Prove that for all $k, n \in \mathbb{N}^+$, with $2k \leqslant n$,

$$s_k(n) = \sum_{t=k}^{n-k} s_t(n-k)$$

(d) Prove that for all $n \in \mathbb{N}^+$,

$$p(n) = s_1(n+1).$$

(e) Prove that for all $k, n \in \mathbb{N}^+$,

$$s_k(n) = s_k(n-k) + s_{k+1}(n+1).$$

3.3 WHAT IS A FORMULA?

Suppose we are given a function which is defined in some way. We have in mind the particular function $p : n \mapsto p(n)$ associated with the partition numbers, but our discussion in this section will apply generally. It is natural to ask whether there is a formula which gives the values of the function. But what exactly does this mean?

It is not difficult to give examples of functions which are defined by formulas. Here are just a few.

$$f_1 : \mathbb{N} \mapsto \mathbb{N}, \text{ given by } f_1 : n \mapsto n(n+1).$$

$$f_2 : \mathbb{N} \mapsto \mathbb{Q}, \text{ given by } f_2 : n \mapsto \frac{n^2+n+1}{n^3+n+1}.$$

$$f_3 : \mathbb{N} \mapsto \mathbb{R}, \text{ given by } f_3 : n \mapsto \cos(n^3).$$

$$f_4 : \mathbb{R} \mapsto \mathbb{N}, \text{ given by } f_4 : x \mapsto \operatorname{int}(e^x).$$

It is more difficult to draw the boundary between what is a formula and what is not. What seems to be clear is that a formula is an expression built up from certain basic symbols according to some rules of syntax which enable us to combine the symbols in various ways. With this approach, whether or not a function can be defined by a formula depends on what basic symbols we are allowed, and how we are allowed to combine them.

Consider, for example, the function p_2 which gives the number of partitions of n into at most two parts. From the proof of Theorem 3.2 we see that

$$p_2(n) = \begin{cases} \dfrac{n+1}{2} & \text{if } n \text{ is odd,} \\[2mm] \dfrac{n}{2} + 1 & \text{if } n \text{ is even.} \end{cases}$$

Is this a formula for the function p_2? Many people would say not, but that instead it is a definition of p_2 which uses two different formulas. However, in the statement of Theorem 3.2 we expressed p_2 by a single formula, namely,

$$p_2(n) = \text{int}\left(\frac{n}{2} + 1\right),$$

making use of the symbol 'int' for the integer-part function. Thus the answer to the question as to whether p_2 can be defined by a formula depends entirely on what symbols we allow in formulas. The split definition above would count as a single formula if we allowed formulas to contain the symbol

$$\{$$

and the second definition of p_2 counts as a formula only if we allow 'int' to occur in formulas.

It now looks as though the answer to the question as to whether a given function can be defined by a formula is entirely arbitrary, depending only on what notation we are willing to allow. If we adopt this point of view, there is a simple answer to the question as to whether the partition number function can be defined by a formula. The answer is 'yes' because

the number of partitions of n is $p(n)$.

Furthermore, this formula is much simpler than in any of the examples given above, since it involves just four symbols, p, '(', n and ')'. What is wrong with this answer?

One answer that might spring to mind is that whereas, for example, the formula for f_2 given earlier in this section, tells us how to calculate the values of f_2, the formula $p(n)$ does not convey any information about how the values of p are to be calculated. However, just as it is a matter of convention that the occurrences of the symbols 'n^3' in the formula for f_2 means 'calculate the cube of n', so equally we could adopt the convention that an occurrence of $p(n)$ in a formula means 'enumerate the partitions of n and count how many you get'. Once we have adopted this convention, the occurrence of $p(n)$ in a formula does tell us what calculation to carry out.

Thus it is just a matter of convention as to whether or not there is a formula for a given function. This suggests that we have been asking the wrong question. What lies behind the question? Why should we wish to know whether or not a particular function can be defined by a formula?

The answer to this question is that there are many natural questions about functions which can be directly answered if we have formulas of certain kinds for them. For example, for real-valued functions of a real variable, that is, for functions $f: \mathbb{R} \mapsto \mathbb{R}$, it is natural to ask whether f is continuous or differentiable. If f can be expressed by a formula involving 'standard' functions, then the answer to these questions can usually be read off from the formula in a very straightforward way.

For a 'counting function', that is, a function which counts the number of arrangements of objects of a certain kind, depending on a parameter n, such as p or p_2, the natural questions to ask include the following.

Is there an algorithm for computing the values of the function?
How efficient is the algorithm, if there is one?
How fast do the values of the function increase?

In this context it becomes sensible to ask whether there is a formula for the values of a function such as p which makes it easy to answer these questions.

There is one class of functions for which these questions are easy to solve. We say that f is a **polynomial function** if f is given by

$$f(n) = a_0 + a_1 n + a_2 n^2 + \ldots + a_k n^k$$

for some $k \in \mathbb{N}$, and $a_0, a_1, \ldots, a_k \in \mathbb{Q}$. (Note that even though we are interested in functions with domain and codomain $\mathbb{N}$, we allow the coefficients to be rational numbers. Thus we are including such functions as $n \mapsto \frac{1}{2}n(n+1)$ which do map natural numbers to natural numbers, even though the coefficients in the polynomial are not all integers. In other contexts it would be appropriate to allow the coefficients to be real numbers, or complex numbers.)

Polynomial functions can be calculated by straightforward algorithms. Furthermore, it is very easy to see from the formula for a polynomial function how many additions and multiplications are needed to calculate its values, giving us a guide to the efficiency of algorithms for computing its values (see Exercise 3.3.1). A polynomial function is dominated by the term of highest degree, so we can also easily tell how rapidly its values increase.

For the function p, there is also a straightforward algorithm for computing its values: systematically enumerate all the partitions of n and count up how many there are. To convince ourselves that this really does yield an algorithm for calculating the values of p, we need to check that there is an algorithm which enables us systematically to list all the partitions of a given natural number n. This problem is considered in Exercise 3.3.2, below.

However, this algorithm for calculating the values of p is not a very efficient one. The number of steps it requires to find the value $p(n)$ is greater than the number $p(n)$ itself. Furthermore, this method yields no significant information about the general behaviour of the function p, for example, how rapidly its values grow.

This discussion leads to the conclusion that the question whether there is a formula for the function p, is not a sensible one to ask. However, it does make sense to ask whether there are formulas for p of particular kinds. Also, one question we should try to settle is whether the function p is a polynomial function. We begin to answer this question in the next section, where we obtain a lower bound for the values of the restricted partition numbers $p_k(n)$.

Exercises

3.3.1. How many additions and multiplications are needed to calculate each of the values of the polynomial function

$$f : n \mapsto a_0 + a_1 n + a_2 n^2 + \ldots + a_k n^k?$$

3.3.2. Show that there is an algorithm for calculating the values of $p(n)$ by enumerating the partitions of n, counting how many there are as you list them. (This is just a matter of showing that there is a systematic method for listing all the partitions of n. One approach would be to write a computer program for the numbers, $A(i, j)$, in a two-dimensional array, such that, for each $n \in \mathbb{N}^+$, the numbers $A(i, j)$ depend on the parameter n and for $1 \leqslant i \leqslant p(n)$,

$$A\ (i, 1) + A(i, 2) + \ldots + A(i, t)$$

is the ith partition of n in the enumeration. Here t is a number, depending on the parameters n and i which specifies the number of terms in the ith partition of n. The enumeration would start with the partition given by

$$A(1, 1) = n$$

and end with the partition given by

$$A(i, j) = 1 \qquad \text{for } 1 \leqslant j \leqslant n.$$

$p(n)$ is then the value of i for this last partition. With this approach all you need do is show how to calculate the values $A(i + 1, j)$ from the values $A(i, j)$. This can be done informally, or by means of a computer program.

(*Warning*: The values of $p(n)$ increase very rapidly—for example $p(50) > 10^5$ and $p(100) > 10^8$. This will lead to problems in implementing your program. First, it is likely to take a long time to run. So do not try to run it for values of n bigger than 50 or so unless you know how to interrupt a program which seems to be running for ever. Second, you will need a great deal of storage space to store the whole array, $A(i, j)$. However, it is not really necessary to do this, as the values, $A(i + 1, j)$, in one array are calculated just from the values in the preceding array,

$A(i, j)$, and not any earlier ones. Hence, all you need to store are the terms, $A(i, j)$, while you are calculating the values, $A(i+1, j)$ in the next partition. The largest j you have to cope with is the largest number of terms that can occur in any partition of n, that is, n itself. More efficient ways of calculating $p(n)$, other than by enumerating all the partitions of n, are mentioned at the end of Chapter 5.)

3.4 A LOWER BOUND FOR $p_k(n)$

We are going to prove, using mathematical induction, an inequality which gives a lower bound for $p_k(n)$. At the induction step we need to use the inequality given in the next lemma. Its relevance may not be immediately apparent, but the form of the inequality should remind you of the formula given in Theorem 3.5.

Lemma 3.6
If $a, b \in \mathbb{R}$ are such that $b > 0$, $a \geqslant 0$, $m \in \mathbb{N}^+$ and $s = \operatorname{int}(a/b)$, then

$$a^{m-1} + (a-b)^{m-1} + (a-2b)^{m-1} + \ldots + (a-sb)^{m-1} > \frac{a^m}{bm}.$$

Proof
Consider the function $f: x \mapsto x^{m-1}$ on the interval $[a-sb, a]$.

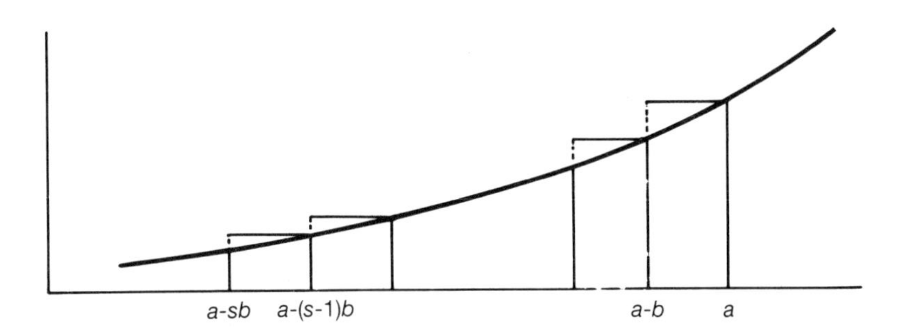

f is increasing on this interval, hence the area under the curve is less than the area of the rectangles shown in the diagram. (More precisely, the integral of f on the interval $[a-sb, b]$ is less than the upper Riemann sum corresponding to the partition $a-sb$, $a-(s-1)b$, $\ldots, a-b, b$ of the interval.) That is

$$\int_{a-sb}^{a} x^{m-1}\, \mathrm{d}x < b(a-(s-1)b)^{m-1} + \ldots + b(a-b)^{m-1} + ba^{m-1},$$

and so

$$\frac{a^m}{m} - \frac{(a - sb)^m}{m} < b(a^{m-1} + (a - b)^{m-1} + \ldots + (a - (s - 1)b)^{m-1}) \tag{3.6}$$

Now, as $s = \text{int}(a/b)$, $0 \leq a - sb < b$, and so

$$(a - sb)^m < b(a - sb)^{m-1}.$$

Since $m \geq 1$, it follows that

$$\frac{(a - sb)^m}{m} < b(a - sb)^{m-1}. \tag{3.7}$$

Adding inequalities (3.6) and (3.7), we obtain

$$\frac{a^m}{m} < b(a^{m-1} + \ldots + (a - sb)^{m-1})$$

from which the inequality of the lemma follows immediately on dividing both sides by b.

We can now use Lemma 3.6 to prove the inequality for $p_k(n)$ that we have been aiming at.

Theorem 3.7
For all $k, n \in \mathbb{N}^+$,

$$p_k(n) \geq \frac{n^{k-1}}{(k - 1)!k!} \tag{3.8}$$

Proof
We prove that expression (3.8) holds for all $k \in \mathbb{N}^+$ by mathematical induction. When $k = 1$, the left-hand side of expression (3.8) is $p_1(n)$ which equals 1, and the right-hand side is $n^0/0!1!$, which also equals 1. Hence the inequality holds for $k = 1$.

Now suppose that $m \in \mathbb{N}^+$, and that the inequality (3.8) holds for $k = m$ and for all $n \in \mathbb{N}^+$. By Theorem 3.5,

$$p_{m+1}(n) = p_m(n) + p_m(n - (m + 1)) + \ldots + p_m(n - s(m + 1)),$$

where $s = \text{int}(n/(m + 1))$. Hence, by our induction hypothesis,

$$p_{m+1}(n) \geq \frac{1}{(m - 1)!m!}(n^{m-1} + (n - (m + 1))^{m-1} + \ldots + (n - s(m + 1))^{m-1}) \tag{3.9}$$

By Lemma 3.6 with $a = n$ and $b = m + 1$, we have

$$n^{m-1} + (n - (m + 1))^{m-1} + \ldots + (n - s(m + 1))^{m-1} > \frac{n^m}{m(m + 1)}. \tag{3.10}$$

From (3.9) and (3.10) it follows that

$$p_{m+1}(n) \geqslant \frac{1}{(m-1)!m!} \cdot \frac{n^m}{m(m+1)} = \frac{n^m}{m!(m+1)!},$$

which shows that the inequality (3.8) holds also for $k = m + 1$. So, by mathematical induction, it follows that this inequality holds for all integers $k \geqslant 1$.

Note that when $n = 2$, Theorem 3.7 gives $p_2(n) \geqslant n/2$, which should be compared with the formula given by Theorem 3.2, that $p_2(n) = \text{int}(n/2 + 1)$. For $n = 3$, it gives $p_3(n) \geqslant n^2/12$, and this inequality should be compared with your answer to Exercise 3.2.1.

We can use Theorem 3.7 to obtain a lower bound for the unrestricted partition numbers, $p(n)$, since $p(n) \geqslant p_k(n)$ for all $k \in \mathbb{N}^+$. All we need do is to choose a suitable value of k in the inequality (3.8) of the theorem. The best choice of k would be the one which makes the right-hand side of the inequality as large as possible. To do this we need to know something about the relationship between the values of n^{k-1} and $k!$. The information that we require is given by Stirling's approximation for $n!$. We make a detour in Chapter 4 to give a proof of Stirling's formula.

4

Stirling's approximation

4.1 ASYMPTOTIC FUNCTIONS

The heart of this chapter is the proof of Stirling's Theorem giving an approximation for $n!$. The proof of this theorem can be found in section 4.2. We then use the approximation to obtain a lower bound for the partition numbers, $p(n)$, as a consequence of the inequality for $p_k(n)$ which we derived in Theorem 3.7.

The proof of Stirling's Theorem uses some elementary ideas from analysis but no combinatorial ideas. Some readers may therefore prefer to take the formula on trust without reading the details of the proof. They should read this section to understand the kind of approximation given by the theorem, and then proceed directly to section 4.4, where we make use of it.

In a polynomial function it is the term of highest degree which is the dominant one. For example, let f be the polynomial function

$$f : n \mapsto n^3 + 5n^2 + 3n + 6.$$

As n gets larger and larger, the value of n^3 dominates the other terms in the sense that the value of $5n^2 + 3n + 6$ becomes more and more negligible in relation to the value of n^3. This can easily be seen from the values in the following table:

n	$n^3 + 5n^2 + 3n + 6$	n^3
10	1 536	1 000
100	1 050 306	1 000 000
1 000	1 005 003 006	1 000 000 000
10 000	1 000 500 030 006	1 000 000 000 000

We can express this by saying that

$$\lim_{n\to\infty} \frac{n^3 + 5n^2 + 3n + 6}{n^3} = 1,$$

or, equivalently, that

$$\lim_{n\to\infty} \frac{5n^2 + 3n + 6}{n^3} = 0.$$

In general, we say that two functions, f and g, are **asymptotic** if

$$\lim_{n\to\infty} \frac{f(n)}{g(n)} = 1 \tag{4.1}$$

or, equivalently, if

$$\lim_{n\to\infty} \frac{f(n) - g(n)}{g(n)} = 0. \tag{4.2}$$

For this definition to make sense we need f and g to be functions whose domain is the set $\mathbb{N}$ of natural numbers, or at any rate, some infinite subset of $\mathbb{N}$. In most combinatorial applications the values of the functions we are interested in will also be natural numbers. However, (4.1) makes sense in a wider context, and since we wish to consider the approximation of integer-valued functions by functions which do not necessarily take only integer values, it is sensible to allow for cases where the codomain of f and g is the set $\mathbb{R}$ of real numbers. Thus our definition of *asymptotic* functions should be taken as applying to functions $f, g : \mathbb{N} \to \mathbb{R}$. (It is possible to define an analogous notion for functions $f, g : \mathbb{R} \to \mathbb{R}$.)

The relation we have just defined is a relation between *functions*. We will write

$$f \sim g$$

to mean that the function f is asymptotic to the function g. When the functions are given by algebraic expressions and we have no convenient short name for them, it is convenient to abuse the notation for functions and to write

$$f(n) \sim g(n)$$

to express the fact that the functions $n \mapsto f(n)$ and $n \mapsto g(n)$ are asymptotic, instead of the strictly correct, but rather cumbersome

$$n \mapsto f(n) \sim n \mapsto g(n).$$

If f is asymptotic to g, the **relative error** in approximating $f(n)$ by $g(n)$ tends to 0 as n increases. This is what equation (4.2) above says. It is important to note that this is compatible with the **absolute error**, $f(n) - g(n)$, becoming unboundedly large. For example, with the particular functions f and g, with which we began this section, although $f \sim g$, $f(n) - g(n) = 5n^2 + 3n + 6$ and so $f(n) - g(n) \to \infty$ as $n \to \infty$.

The relation $\sim$ is an equivalence relation between functions. (You are asked to check this in Exercise 4.1.1.) Two polynomials are in the same equivalence class if and only if their dominant terms are the same. A polynomial function can very easily also be asymptotic to a function which is not a polynomial function. For example, we have

$$n^3 \sim n^3 + \ln(n).$$

It is easy to prove that if a function is asymptotic to a polynomial function, then even if it is not a polynomial function itself, it is bounded by a polynomial function.

Lemma 4.1

If the function f is asymptotic to a polynomial function of degree k, then there is some constant $A > 0$ such that for all $n \in \mathbb{N}$,

$$|f(n)| < An^k. \tag{4.3}$$

Proof

Suppose that $f \sim g$ where $g : n \mapsto a_k n^k + \ldots + a_1 n + a_0$ with $a_k \neq 0$. Then $g \sim h$, where $h : n \mapsto a_k n^k$, and so $f \sim h$. Hence

$$\lim_{n \to \infty} \frac{f(n)}{a_k n^k} = 1.$$

Since the sequence

$$\left\{ \frac{f(n)}{a_k n^k} \right\}$$

has a limit, by a basic theorem of analysis, it is bounded, that is, there is some constant $C > 0$ such that for all $n \in \mathbb{N}$,

$$\left| \frac{f(n)}{a_k n^k} \right| < C. \tag{4.4}$$

The inequality (4.3) of the lemma follows from inequality (4.4) on putting

$$A = |a_k| C.$$

We say that a function f is **dominated by a polynomial function** if there is a polynomial function g such that for all $n \in \mathbb{N}$,

$$|f(n)| \leq g(n).$$

Thus the lemma above says that if a function f is asymptotic to a polynomial function then it is dominated by a polynomial function. The converse need not be true. For example, the function f given by

$$f(n) = \begin{cases} n^2, & \text{if } n \text{ is even,} \\ n^3, & \text{if } n \text{ is odd.} \end{cases}$$

is dominated by the polynomial function $n \mapsto n^3$, but is not asymptotic to any polynomial function.

Our definition of a function, f, being dominated by a polynomial function, g, requires that the inequality, $|f(n)| \leqslant g(n)$, should hold for *all* $n \in \mathbb{N}$. As is shown by Exercise 4.1.4, it makes no difference if we allow a finite number of exceptions, so that $|f(n)| \leqslant g(n)$ holds only for all natural numbers, n, which are larger than some fixed number n_0.

In section 3.3, we suggested that a natural question to ask about a function such as p, is how fast the function grows. If we knew that p was asymptotic to a polynomial function, or even dominated by one, this would give a reasonably precise answer to our question. However, these possibilities can easily be ruled out by considering the inequality of Theorem 3.7.

For suppose p is dominated by a polynomial function and that for all $n \in \mathbb{N}$,

$$p(n) \leqslant An^k \tag{4.5}$$

for some $k \in \mathbb{N}^+$, and some constant $A \in \mathbb{R}$. Replacing k by $k+2$ in Theorem 3.7, we have that, for all $n \in \mathbb{N}$,

$$\frac{n^{k+1}}{(k+1)!(k+2)!} \leqslant p_{k+2}(n) \leqslant p(n). \tag{4.6}$$

By inequalities (4.5) and (4.6), for all $n \in \mathbb{N}$,

$$\frac{n^{k+1}}{(k+1)!(k+2)!} \leqslant An^k,$$

which is impossible as it implies that for all $n \in \mathbb{N}$,

$$n \leqslant A(k+1)!(k+2)!$$

Thus we have proved the following theorem:

Theorem 4.2

The function p is not dominated by a polynomial function and hence is not asymptotic to a polynomial function.

Theorem 4.2, by itself, tells us that the values of p must grow very rapidly. For example it tells us that there are infinitely many values of $n \in \mathbb{N}$ such that

$$p(n) > n^{1\,000\,000},$$

and we could replace the right-hand side of this inequality by any power of n we choose. After we have proved Stirling's Theorem in the next section, we shall be able to derive more information about the rapidity of growth of the function p.

Exercises

4.1.1. I would like you to prove that the relation $\sim$ is an equivalence relation on the set of all functions $f, f: \mathbb{N} \to \mathbb{R}$. Unfortunately, there is a small technical difficulty which means that this is not quite true. For if f is a function such that $\{n \in \mathbb{N} : f(n) = 0\}$ is infinite, then the limit

$$\lim_{n\to\infty} \frac{f(n)}{f(n)}$$

does not exist, and so cannot equal 1. Thus for such functions, f is not asymptotic to itself, showing that the relation $\sim$ is not reflexive. We can easily get over this minor problem by stipulating that even in such a case, $f \sim f$.

Now prove that $\sim$ is indeed an equivalence relation on the set of all functions $f, f: \mathbb{N} \to \mathbb{R}$.

4.1.2. Prove that if f, g are functions such that $f(n) \sim g(n)$, and g never takes the value 0, then there is some constant C such that for all $n \in \mathbb{N}$,

$$|f(n)| < C|g(n)|.$$

4.1.3. Prove that, if $a_k \neq 0$, then

$$a_k n^k + \ldots + a_1 n + a_0 \sim a_k n^k.$$

Deduce that two polynomial functions

$$f: n \mapsto a_j n^j + \ldots + a_1 n + a \text{ and } g \mapsto b_k n^k + \ldots + b_1 n + b_0,$$

with $a_j \neq 0$, and $b_k \neq 0$, are asymptotic if and only if $j = k$ and $a_j = b_k$.

4.1.4. Prove that for each function $f: \mathbb{N} \to \mathbb{R}$, if there is a polynomial function g and an integer n_0 such that for all $n \in \mathbb{N}$ with $n \geqslant n_0$,

$$|f(n)| \leqslant g(n),$$

then f is dominated by some polynomial function.

4.2 STIRLING'S FORMULA

In this section we prove Stirling's asymptotic formula for $n!$, namely that

$$n! \sim \left(\frac{n}{\mathrm{e}}\right)^n \sqrt{2\pi n}$$

where, of course, e is the base of natural logarithms, i.e. $\mathrm{e} \simeq 2.71828$. This is equivalent to proving that

$$\lim_{n\to\infty} \frac{\left(\dfrac{n}{\mathrm{e}}\right)^n \sqrt{n}}{n!} = \frac{1}{\sqrt{2\pi}}. \tag{4.7}$$

Our strategy in proving equation (4.7) will be to first show that the limit on the left-hand side exists, and then to show that the value of this limit is indeed $1/\sqrt{2\pi}$. We proceed by a sequence of lemmas. Both $n!$ and $(n/\mathrm{e})^n$ involve products, whereas most of the basic results about limits are stated in terms of limits of sums (for example, those relating to the convergence of series). Our first step therefore involves replacing the products on the left-hand side of equation (4.7) by a sum, by using logarithms.

Lemma 4.3
For all $n \in \mathbb{N}$ with $n \geqslant 2$,

$$\ln\left(\frac{\left(\dfrac{n}{\mathrm{e}}\right)^n \sqrt{n}}{n!}\right) = \sum_{k=1}^{n-1}\left(\left(k+\frac{1}{2}\right)\ln\left(\frac{k+1}{k}\right) - 1\right) - 1 \tag{4.8}$$

Proof
We prove equation (4.8) by mathematical induction. For $n = 2$, the left-hand side of equation (4.8) is

$$\ln\left(\frac{\left(\dfrac{2}{\mathrm{e}}\right)^2 \sqrt{2}}{2!}\right) = \ln\left(\frac{2\sqrt{2}}{\mathrm{e}^2}\right) = \frac{3}{2}\ln 2 - 2 = \left(\frac{3}{2}\ln\frac{2}{1} - 1\right) - 1,$$

which is also the value of the right-hand side for $n = 2$. Thus equation (4.8) holds in this case.

Now assume, as our induction hypothesis, that equation (4.8) holds when $n = m$. Then, we have

$$\begin{aligned}
&\sum_{k=1}^{m}\left(\left(k+\frac{1}{2}\right)\ln\left(\frac{k+1}{k}\right) - 1\right) - 1 \\
&\quad= \sum_{k=1}^{m-1}\left(\left(k+\frac{1}{2}\right)\ln\left(\frac{k+1}{k}\right) - 1\right) - 1 + \left(\left(m+\frac{1}{2}\right)\ln\left(\frac{m+1}{m}\right) - 1\right) \\
&\quad= \ln\left(\frac{\left(\dfrac{m}{\mathrm{e}}\right)^m \sqrt{m}}{m!}\right) + \left(\left(m+\frac{1}{2}\right)\ln\left(\frac{m+1}{m}\right) - 1\right),
\end{aligned}$$

using our induction hypothesis that equation (4.8) holds when $n = m$,

$$= \ln\left(\frac{\left(\frac{m}{e}\right)^m \sqrt{m}}{m!}\right) + \ln\left(\frac{m+1}{m}\right)^{m+1/2} + -(\ln(e))$$

$$= \ln\left(\frac{\left(\frac{m}{e}\right)^m \sqrt{m}}{m!} \times \left(\frac{m+1}{m}\right)^{m+1/2} \times \frac{1}{e}\right)$$

$$= \ln\left(\frac{\left(\frac{m+1}{e}\right)^{m+1} \sqrt{m+1}}{(m+1)!}\right).$$

Hence equation (4.8) holds also when $n = m + 1$. Therefore, by mathematical induction, it holds for all $n \geqslant 2$.

It follows that in order to prove that the limit in equation (4.7) exists, it is sufficient to prove that the series whose partial sums occur on the right-hand side of equation (4.8) converges. We do this in the next lemma.

Lemma 4.4
The series

$$\sum_{k=1}^{\infty}\left(\left(k + \frac{1}{2}\right)\ln\left(\frac{k+1}{k}\right) - 1\right)$$

converges.

Proof
We make use of the standard power series for the functions $x \mapsto \ln(1 + x)$ and $x \mapsto \ln(1 - x)$. These tell us that, for $0 \leqslant x < 1$,

$$\ln(1 + x) = x - \frac{x^2}{2} + \frac{x^3}{3} - \ldots \tag{4.9}$$

and

$$\ln(1 - x) = -x - \frac{x^2}{2} - \frac{x^3}{3} - \ldots \tag{4.10}$$

Subtracting equation (4.10) from (4.9) and dividing by 2, we obtain

$$\frac{1}{2}\ln\left(\frac{1+x}{1-x}\right) = x + \frac{x^3}{3} + \frac{x^5}{5} + \ldots$$

Hence

$$x \leqslant \frac{1}{2}\ln\left(\frac{1+x}{1-x}\right) \leqslant x + \frac{x^3}{3} + \frac{x^5}{3} + \ldots$$

$$= x + \frac{x^3}{3}(1 + x^2 + x^4 + \ldots)$$

$$= x + \frac{x^3}{3(1-x^2)}, \tag{4.11}$$

making use of the fact that as $|x| < 1$, the series $(1 + x^2 + x^4 + \ldots)$ converges with sum $1/(1-x^2)$.

We now put $x = 1/(2k+1)$, for $k \in \mathbb{N}^+$. It follows that $0 \leqslant x < 1$, and it is easy to see that

$$\frac{1+x}{1-x} = \frac{k+1}{k}$$

and

$$\frac{x^3}{3(1-x^2)} = \frac{1}{12(k^2+k)(2k+1)}.$$

Hence, from equation (4.11) we have

$$\frac{1}{2k+1} \leqslant \frac{1}{2}\ln\left(\frac{k+1}{k}\right) \leqslant \frac{1}{2k+1} + \frac{1}{12(k^2+k)(2k+1)}$$

and hence, on multiplying through by $(2k+1)$ and rearranging, we get

$$0 \leqslant \left(k + \frac{1}{2}\right)\ln\left(\frac{k+1}{k}\right) - 1 \leqslant \frac{1}{12(k^2+k)} < \frac{1}{k^2}. \tag{4.12}$$

Now

$$\sum_{k=1}^{\infty} \frac{1}{k^2}$$

is a standard convergent series, and therefore it follows from (4.12) and the Comparison Test for series of non-negative terms that the series of the lemma converges.

As we have already noted, it follows from Lemmas 4.3 and 4.4 that

$$\lim_{n\to\infty} \ln\left(\frac{\left(\frac{n}{e}\right)^n \sqrt{n}}{n!}\right)$$

exists, and hence that

$$\lim_{n\to\infty} \frac{\left(\frac{n}{e}\right)^n \sqrt{n}}{n!}$$

exists. We now set about proving that the value of this limit is $1/\sqrt{(2\pi)}$.

Our starting point is the integral

$$I_n = \int_0^{\pi/2} \sin^n x \, dx.$$

It is a straightforward exercise in integration by parts to show that for each $n \in \mathbb{N}^+$,

$$I_n = \frac{n-1}{n} I_{n-2},$$

and hence that

$$I_{2n} = \frac{(2n)!\pi}{2^{2n+1}(n!)^2}$$

$$I_{2n+1} = \frac{2^{2n}(n!)^2}{(2n+1)!}. \tag{4.13}$$

Since for $0 \leqslant x \leqslant \pi/2$, we have $0 \leqslant \sin x \leqslant 1$, it follows that for all $n \in \mathbb{N}^+$,

$$\sin^{2n+1} x \leqslant \sin^{2n} x \leqslant \sin^{2n-1} x,$$

and hence that

$$I_{2n+1} \leqslant I_{2n} \leqslant I_{2n-1}.$$

Therefore, by (4.13)

$$\frac{2^{2n}(n!)^2}{(2n+1)!} \leqslant \frac{(2n)!\pi}{2^{2n+1}(n!)^2} \leqslant \frac{2^{2n-2}((n-1)!)^2}{(2n-1)!}. \tag{4.14}$$

From the first inequality in expression (4.14) it follows that

$$\frac{2^{4n}(n!)^4}{(2n)!(2n+1)!} \leqslant \frac{\pi}{2} \tag{4.15}$$

and from the second, that

$$\frac{n\pi}{(2n+1)} \leqslant \frac{2^{4n}(n!)^4}{(2n)!(2n+1)!}. \tag{4.16}$$

Now

$$\lim_{n\to\infty} \frac{n\pi}{(2n+1)} = \frac{\pi}{2}$$

and hence, it follows from inequalities (4.15) and (4.16) by the Squeeze

Rule for limits that

$$\lim_{n\to\infty} \frac{2^{4n}(n!)^4}{(2n)!(2n+1)!} = \frac{\pi}{2}. \tag{4.17}$$

Although it may not yet seem so, the evaluation of this limit is a major step towards the evaluation of the limit we are really interested in. We can put

$$a = \lim_{n\to\infty} \frac{\left(\dfrac{n}{\mathrm{e}}\right)^n \sqrt{n}}{n!}$$

now that we know that the limit exists. It follows that also

$$a = \lim_{n\to\infty} \frac{\left(\dfrac{2n}{\mathrm{e}}\right)^{2n} \sqrt{2n}}{(2n)!}$$

and hence that

$$\frac{1}{a^2} = \frac{a^2}{a^4} = \frac{\left(\lim\limits_{n\to\infty} \dfrac{(2n/\mathrm{e})^{2n}\sqrt{2n}}{(2n)!}\right)^2}{\left(\lim\limits_{n\to\infty} \dfrac{(n/\mathrm{e})^n\sqrt{n}}{n!}\right)^4}$$

$$= \lim_{n\to\infty} \frac{\dfrac{2^{4n}n^{4n}2n}{\mathrm{e}^{4n}((2n)!)^2}}{\dfrac{n^{4n}n^2}{\mathrm{e}^{4n}(n!)^4}},$$

using the Product and Quotient Rules for limits,

$$= \lim_{n\to\infty} \frac{2^{4n+1}(n!)^4}{n((2n)!)^2}$$

$$= \lim_{n\to\infty} \left(\frac{2^{4n}(n!)^4}{(2n)!(2n+1)!} \times \frac{2(2n+1)}{n}\right)$$

$$= \frac{\pi}{2} \times 4 = 2\pi,$$

using equation (4.17) and the Product Rule for limits. It therefore follows that $a = 1/\sqrt{2\pi}$. Thus, at long last, we have proved the following:

Theorem 4.5 *(Stirling's approximation for n!)*

$$n! \sim \left(\frac{n}{\mathrm{e}}\right)^n \sqrt{2\pi n}.$$

The advantages of Stirling's approximation will become apparent when we make use of it in section 4.4. Other applications are given in Exercises 4.2.2 and 4.2.3. If we rewrite the formula by taking logarithms, we get

$$\ln(n!) \sim n(\ln(n) - 1) + \tfrac{1}{2}\ln(2\pi n),$$

which gives us a very quick way to calculate good approximations to $\ln(n!)$.

It is worth noting, in passing, that the limit of equation (4.17) can be rewritten in terms of an infinite product. The numerator $2^{4n}(n!)^4$, on the left-hand side of equation (4.17), can be written as $(2^n n!)^4$, and hence the whole expression equals

$$\frac{(2 \times 4 \times 6 \times \ldots \times 2n)^4}{(1 \times 2 \times 3 \times \ldots \times 2n)(1 \times 2 \times 3 \times \ldots \times (2n+1))}$$

$$= \frac{2 \times 2 \times 4 \times 4 \times \ldots \times 2n \qquad \times 2n}{1 \times 3 \times 3 \times 5 \times \ldots \times 2n - 1 \times 2n + 1}$$

Hence it follows from equation (4.17) that

$$\frac{2 \times 2 \times 4 \times 4 \times \ldots \times 2n \qquad \times 2n \times \ldots}{1 \times 3 \times 3 \times 5 \times \ldots \times 2n - 1 \times 2n + 1 \times \ldots} = \frac{\pi}{2} \tag{4.18}$$

Equation (4.18) is known as **Wallis's formula** for $\pi/2$. Although the infinite product does converge to $\pi/2$, it does so rather slowly. For example, with $n = 10$ we get the value 1.53385 compared with the true value of 1.57080 for $\pi/2$ (both values given to five decimal places).

The approximation for $n!$ given by Stirling's formula is accurate enough for most purposes. We can improve it by taking more care of the inequalities. In this way it is possible to improve Stirling's formula to

$$n! \sim \left(1 + \frac{1}{12n}\right)\left(\frac{n}{\mathrm{e}}\right)^n \sqrt{2\pi n}$$

and further refinements are possible. We will content ourselves with the approximation given by Theorem 4.5, as this gives the value of $n!$ with an error of less than 1% for all $n \geqslant 10$. Table 4.1 compares the values given by these approximations with the true value of $n!$ for selected values of n.

We are now ready to return to the partition numbers, and to use what we have learned about the values of $n!$ to derive a lower bound for $p(n)$. This we do in section 4.4.

Table 4.1 Approximations of $n!$ given by Stirling's formula

n	$n!$	$\left(\frac{n}{e}\right)^n\sqrt{2\pi n}$	$\left(1+\frac{1}{12n}\right)\left(\frac{n}{e}\right)^n\sqrt{2\pi n}$
1	1	0.922	0.999
2	2	1.919	1.999
3	6	5.836	5.998
4	24	23.506	23.996
5	120	118.019	119.986
6	720	710.078	719.940
7	5 040	4 980.396	5 039.686
8	40 320	39 902.395	40 318.045
9	362 880	359 536.873	362 865.918
10	3 628 800	3 598 695.619	3 628 684.749
20	2.433×10^{18}	2.423×10^{18}	2.433×10^{18}
30	2.653×10^{32}	2.645×10^{32}	2.653×10^{32}
40	8.159×10^{47}	8.142×10^{47}	8.159×10^{47}
50	3.041×10^{64}	3.036×10^{64}	3.041×10^{64}
60	8.321×10^{81}	8.309×10^{81}	8.321×10^{81}

Exercises

4.2.1. Prove that the formulas for I_{2n} and I_{2n+1} given in equations (4.13) are correct.

4.2.2. Find the least positive integer n such that

$$n! > 10^{10^6}.$$

4.2.3. Let e_n be the probability that if a fair coin is tossed $2n$ times, the result is n heads and n tails. Prove that

$$e_n \sim \frac{1}{\sqrt{\pi n}}.$$

4.3 A NOTE ON JAMES STIRLING

James Stirling was born in Scotland in 1692. This was only four years after the Stuart King, James II, had been forced to abdicate, and was replaced by the Hanoverian King, George I. James II tried unsuccessfully to regain his throne, and died in exile in 1701. His son, James, the 'Old Pretender', invaded Britain in 1715 in an attempt to regain the throne for the Stuarts. There was another unsuccessful attempt when James II's grandson, Charles Edward, 'Bonny Prince Charles, the Young Pretender', invaded Britain in 1745. He narrowly escaped capture after the battle of Culloden in 1746.

These events formed the backcloth to Stirling's life. His family were Jacobites, that is, supporters of the Stuarts after James II was deposed.

Stirling went first as a student to Glasgow University and subsequently to Oxford, which he managed to enter as a student without taking an oath of allegiance to King George. In Oxford he was acquitted on a charge of cursing the King, but he left without taking a degree.

From Oxford Stirling went to Venice. He settled in London in 1724. His major mathematical work, which includes his approximation for $n!$, *Methodus differentialis: sive tractatus de summatione et interpolatione serierum infinitarum*, was published in 1730. An English translation was published in 1749. For political reasons he was unable to succeed to Colin Maclaurin's chair at Edinburgh University.

In his later life Stirling turned his attention to engineering. He was employed by the Scottish Mining Company and moved back to Scotland in 1735 to Leadhills, Lanarkshire, a village near the company's lead mines. He died in Edinburgh on 5 December 1770.

4.4 A LOWER BOUND FOR $p(n)$

Remember that the digression in section 4.2 was made because we wanted to use the inequality

$$p_k(n) \geqslant \frac{n^{k-1}}{k!(k-1)!} \tag{4.19}$$

of Theorem 3.7 to obtain a lower bound for $p(n)$ by making an appropriate choice of k in this inequality.

It follows from Stirling's approximation, and Exercise 4.1.2, that there is a constant C such that for $n \in \mathbb{N}$,

$$n! < C\left(\frac{n}{\mathrm{e}}\right)^n \sqrt{2\pi n} \tag{4.20}$$

and, in fact, we could take C to be equal to 1.1 in this inequality. It follows from inequality (4.20) that

$$\begin{aligned}\frac{n^{k-1}}{k!(k-1)!} &= \frac{kn^{k-1}}{(k!)^2}\\ &\geqslant \frac{kn^{k-1}}{C^2\left(\frac{k}{\mathrm{e}}\right)^{2k} 2\pi k}\\ &= \frac{n^{k-1}\mathrm{e}^{2k}}{2\pi C^2 k^{2k}}\end{aligned}$$

Since, for all $k \in \mathbb{N}$, $p(n) \geqslant p_k(n)$, it therefore follows from inequality (4.19) that for all $k \in \mathbb{N}$,

$$p(n) \geqslant \frac{n^{k-1}\mathrm{e}^{2k}}{2\pi C^2 k^{2k}} = \left(\frac{1}{2\pi n C^2}\right)\frac{n^k \mathrm{e}^{2k}}{k^{2k}} \tag{4.21}$$

We now choose k so as to make $n^k e^{2k}/k^{2k}$ as large as possible. It can easily be checked, using the methods of elementary calculus, that, for $x > 0$ the function

$$x \mapsto \frac{n^x e^{2x}}{x^{2x}}$$

has a maximum when $x = \sqrt{n}$. As k has to be an integer in inequality (4.21), we put

$$k = \text{int}(\sqrt{n}).$$

It follows that

$$\sqrt{n} - 1 < k \qquad \text{and} \qquad k^2 \leqslant n$$

and hence

$$\frac{n^k e^{2k}}{k^{2k}} > \frac{(k^2)^k e^{2(\sqrt{n}-1)}}{k^{2k}} = \frac{e^{2\sqrt{n}}}{e^2}$$

and so, by inequality (4.21), for all $n \in \mathbb{N}$,

$$p(n) > \frac{1}{2\pi C^2 e^2}\left(\frac{e^{2\sqrt{n}}}{n}\right) \tag{4.22}$$

This inequality for $p(n)$ has been obtained by elementary methods, and so it is not suprising that it is rather crude. Although it does not provide us with an asymptotic formula for $p(n)$, the basic form of the inequality is correct. Using more sophisticated methods it can be proved that

$$p(n) \sim \frac{1}{4\sqrt{3}}\left(\frac{e^{\pi\sqrt{(2/3)n}}}{n}\right). \tag{4.23}$$

This formula is contained in a celebrated paper published by G. H. Hardy and S. Ramanujan in 1918. We discuss this briefly at the end of Chapter 5.

The expressions on the right-hand sides of expressions (4.22) and (4.23) both have the form

$$\frac{\alpha e^{\beta\sqrt{n}}}{n},$$

where α and β are constants. It is the value of the constant β which has the dominant effect on the growth of the values of the corresponding function. In the asymptotic formula (4.23), the value of this constant (to 3 decimal places) is

$$\pi\sqrt{\tfrac{2}{3}} = 2.565$$

as compared with the constant 2 in the corresponding place in expression (4.23). So although inequality (4.22) does not give the correct asymptotic value for this constant, it is not too far out.

Exercises

4.4.1. Verify that $x = \sqrt{n}$ does give a maximum value of the function

$$x \mapsto \frac{n^x e^{2x}}{x^{2x}} \qquad \text{for } x > 0.$$

4.4.2. Compare the values given by the expressions on the right-hand sides of expressions (4.22), with $C = 1.1$, and (4.23) and the value of $p(n)$ for $n = 1, 2, \ldots, 10$.

5

Partitions and generating functions

5.1 INTRODUCTION

In this chapter we describe a powerful algebraic technique which can be used to solve many combinatorial problems. Our concern in this chapter is with the application of this technique to the partition numbers of Chapter 3. In the next chapter we show how this same technique can be applied to other combinatorial problems.

We saw a connection between algebra and counting in Chapter 1 where we gave a combinatorial proof of the Binomial Theorem. Recall that we did this by considering the terms that are obtained when the product

$$(a + b)(a + b) \dots (a + b)$$

is multiplied out in full. We now apply a similar idea to partitions. The following problem, although very simple, will set us off in the right direction.

Problem 1

What is the coefficient of x^7 in the product

$$(1 + x + x^2 + x^3 + \dots)(1 + x^2 + x^4 + x^6 + \dots)? \qquad (5.1)$$

Solution

The coefficient of x^7 is 4 because when the terms in the first bracket are multiplied by the terms in the second bracket we obtain just 4 terms of degree 7, namely,

$$x^1 \cdot x^6, \quad x^3 \cdot x^4, \quad x^5 \cdot x^2 \quad \text{and} \quad x^7 \cdot x^0.$$

We can emphasize the fact that the terms in the first bracket are multiples of 1, and those in the second multiples of 2, by rewriting these four terms as

$$x^{1\cdot 1}\cdot x^{3\cdot 2},\quad x^{3\cdot 1}\cdot x^{2\cdot 2},\quad x^{5\cdot 1}\cdot x^{1\cdot 2}\quad \text{and}\quad x^{7\cdot 1}\cdot x^{0\cdot 2}.$$

Thus we see that these four terms correspond to the solutions of the equation

$$m_1\cdot 1 + m_2\cdot 2 = 7, \qquad \text{with } m_1, m_2 \in \mathbb{N}. \tag{5.2}$$

We can rewrite equation (5.2) as

$$\underbrace{(2 + 2 + \ldots + 2)}_{m_2 \text{ terms}} + \underbrace{(1 + 1 + \ldots + 1)}_{m_1 \text{ terms}} = 7$$

from which we see that the solutions of equation (5.2) correspond to partitions of n into parts of size at most 2. In Chapter 4 we used the notation $q_k(n)$ for the number of partitions of n into parts of size at most k. Thus the fact that the coefficient of x^7 in the expression (5.1) is 4 corresponds to the fact that $q_2(7) = 4$.

We can generalize this observation. We see that the coefficient of x^n in the product (5.1) is $q_2(n)$. Thus we can write

$$(1 + x + x^2 + x^3 + \ldots)(1 + x^2 + x^4 + x^6 + \ldots) = \sum_{n=0}^{\infty} q_2(n)x^n.$$

This will further generalize to deal with partitions of size at most k as follows.

Theorem 5.1

For each $k \in \mathbb{N}^+$,

$$(1 + x + x^2 + \ldots)(1 + x^2 + x^4 + \ldots) \ldots (1 + x^k + x^{2k} + \ldots) = \sum_{n=0}^{\infty} q_k(n)x^n \tag{5.3}$$

Proof

When the product on the left-hand side of equation (5.3) is expanded the terms that we get are obtained by taking one term from each bracket and multiplying them together. Thus these terms have the form

$$x^{m_1\cdot 1}\cdot x^{m_2\cdot 2} \ldots x^{m_k\cdot k}, \qquad \text{with } m_1, \ldots, m_k \in \mathbb{N}$$

A term of this form has degree n if and only if

$$m_1\cdot 1 + m_2\cdot 2 + \ldots + m_k\cdot k = n \tag{5.4}$$

and so the number of terms of degree n is the same as the number of solutions of equation (5.4) with $m_1, \ldots, m_k \in \mathbb{N}$. A solution of equation

(5.4) corresponds to a partition of n with m_1 parts of size 1, m_2 parts of size 2, up to and including m_k parts of size k. That is, a solution of equation (5.4) corresponds to a partition of n into parts of size at most k. So the coefficient of x^n is the same as the number of such partitions of n.

Equation (5.3) involves the sums of infinite series, and we have not yet said how these are to be interpreted. In fact we can view the matter in two ways, *algebraic* and *functional*. We could regard equations such as (5.3) as dealing with formal algebraic expressions. We adopted this viewpoint in giving the proof of Theorem 5.1. The awkwardness of this approach arises when it comes to saying exactly what we mean by a *formal algebraic expression*. The coefficients in the series look as though they are numbers, but what sort of animal is x?

It is tempting to say that x is a symbol, but then how can we combine a symbol with an abstract object like a number to form such things as $3x$? Older writers used to talk about 'an indeterminate x' without ever really saying what this means. The modern approach, which aims to explain all mathematical concepts ultimately in terms of set theory, is to regard an algebraic expression such as

$$a_0 + a_1x + a_2x^2 + \ldots$$

as simply a convenient way of representing the sequence of numbers

$$(a_0, a_1, a_2, \ldots)$$

which is to be manipulated according to certain rules, which arise from the algebraic background to this approach. (The sequences are infinite if we are dealing with power series, and finite if we are dealing with polynomials. There is, of course, no difficulty in defining what we mean by an infinite sequence in terms of sets. If X is a set, an infinite sequence of elements of X can be regarded as a function $f: \mathbb{N} \to X$.)

In this approach the rule for addition becomes

$$(a_0, a_1, a_2, \ldots) + (b_0, b_1, b_2, \ldots) = (a_0 + b_0, a_1 + b_1, a_2 + b_2, \ldots) \tag{5.5}$$

corresponding to term-by-term addition of power series or polynomials. The multiplication rule is a little more complicated. If we multiply the terms in the infinite series $a_0 + a_1x + a_2x^2 + \ldots$ by those in $b_0 + b_1x + b_2x^2 + \ldots$ and gather together terms of the same degree, we see that the coefficient of x^n is

$$a_0b_n + a_1b_{n-1} + \ldots + a_nb_0$$

which we can write as

$$\sum_{i=0}^{n} a_ib_{n-i}.$$

Thus the multiplication rule is

$$(a_0, a_1, a_2, \ldots) \times (b_0, b_1, b_2, \ldots) = (c_0, c_1, c_2, \ldots)$$

where

$$c_n = \sum_{i=0}^{n} a_i b_{n-i}. \tag{5.6}$$

The formulas (5.5) and (5.6) define the algebra of formal power series without any need to mention x. None the less we shall continue to manipulate power series using the traditional notation involving powers of x, just as, even after complex numbers are explained as being pairs, (x, y), of real numbers, it is still convenient to use the traditional notation $x + \mathrm{i}y$ when doing calculations with them.

The second approach to equation (5.3) is to regard it as an equation between *functions*. From this point of view it says that, for a certain range of values of x, the function defined by the formula on the left-hand side has the same values as the function which is defined by the formula on the right-hand side of the equation. Since, in the case of equations like (5.4) the functions are defined by infinite series, this approach involves knowing something about their convergence. Fortunately, it is usually not important to know the exact range of values of x for which the series converge. What matters is that they should converge for at least some non-zero values of x. In other words, we require that the series have a positive radius of convergence, and from time to time we shall quote the standard theorems from analysis which provide us with this information.

There is no difficulty with the particular power series that occur in equation (5.3), since they are all geometric series. Standard theorems thus tell us that they converge for $|x| < 1$, and hence that the product series, as defined by equation (5.6), also converges for this range of values of x.

These two approaches, algebraic and functional, are closely related. Another standard theorem tells us that if the series $\Sigma_{n=0}^{\infty} a_n x^n$ and $\Sigma_{n=0}^{\infty} b_n x^n$ both converge for $|x| < R$, with $R > 0$, then

$$\text{for all } n \in \mathbb{N},\ a_n = b_n \Leftrightarrow \text{for } |x| < R,\ \sum_{n=0}^{\infty} a_n x^n = \sum_{n=0}^{\infty} b_n x^n \tag{5.7}$$

This tells us that an equation such as (5.3) is true when regarded as an algebraic equation between formal power series if and only if it is also true when thought of as a functional equation. It is this interplay which enables us to use both algebraic and analytic methods as appropriate and which turns out to be so fruitful. For example, we often use condition (5.7) to argue that if two functions are the same, then the coefficients in the power series which represent the functions must match exactly. We call this the **method of equating coefficients**.

You may think that rather too much fuss has been made about the two different interpretations of equation (5.3), especially now that we have seen that they are equivalent. However, it is important to be aware that the equivalence (5.7) does depend on both series having a positive radius of convergence. It also depends on properties of the field of real numbers. The analogous result for *finite* fields is not true. For example, the polynomials $1 + x^2 + x^3$ and $1 + x + x^2$ are distinct as polynomials, but they define exactly the same functions on the three-element field $\mathbb{Z}_3$.

It is convenient at this point to introduce some notation for sequences. We write

$$\{a_n\}_{n=0}^{\infty}$$

to represent the infinite sequence

$$(a_0, a_1, \ldots)$$

and we write

$$\{a_n\}_{n=i}^{j}$$

(for $i \leqslant j$) to represent the finite sequence

$$(a_i, a_{i+1}, \ldots, a_{j-1}, a_j).$$

It is convenient to adopt the convention that when the infinite sequence is intended, the part of this notation which specifies the range of values of n can be omitted, so that we can simply write $\{a_n\}$ for this sequence. Sometimes we extend this convention to other cases where the range of values taken by n can be understood from the context.

5.2 GENERATING FUNCTIONS

If we have a sequence $\{a_n\}$ of numbers, the function

$$x \mapsto \sum_{n=0}^{\infty} a_n x^n$$

is called the **generating function** of the sequence. We will only use this terminology when the power series has a positive radius of convergence, so that the method of equating coefficients is applicable. In the combinatorial applications that we consider $\{a_n\}$ will almost always turn out to be a sequence of natural numbers, since a_n will be the number of arrangements of some kind, but the notion of a generating function has wider applications. Hence there is no reason why we should not consider generating functions corresponding to sequences of real numbers or of complex numbers.

We begin our discussion of generating functions by listing some standard examples.

Example 1

Another way of stating the Binomial Theorem is to say that the function $x \mapsto (1+x)^n$ is the generating function for the sequence

$$\{C(n, r)\}_{r=0}^{n}$$

of binomial coefficients. Since the sequence is finite, the generating function is a polynomial function and questions of convergence do not arise.

Example 2

For $|x| < 1$, the power series $\Sigma_{n=0}^{\infty} x^n$ converges and has sum $1/(1-x)$. Thus the generating function for the constant sequence $\{1\}$ is the function $x \mapsto 1/(1-x)$.

Example 3

A theorem of analysis tells us that a power series may be differentiated term by term within its interval of convergence. Hence by differentiating the power series of Example 2, we can deduce that for $|x| < 1$,

$$\frac{1}{(1-x)^2} = \sum_{n=0}^{\infty} nx^{n-1},$$

and hence that

$$\frac{x}{(1-x)^2} = \sum_{n=0}^{\infty} nx^n.$$

Thus the generating function for the sequence $\{n\}$ is

$$x \mapsto \frac{x}{(1-x)^2}.$$

Example 4

Let k be some fixed positive integer. For $|x| < 1$, the geometric series $\Sigma_{n=0}^{\infty} (x^k)^n$ converges and has sum $1/(1-x^k)$. Hence it follows from Theorem 5.1 that the generating function for the sequence $\{q_k(n)\}$ is

$$x \mapsto \frac{1}{(1-x)(1-x^2)\ldots(1-x^k)}.$$

Hence, by Theorem 3.3, this is also the generating function for the sequence $\{p_k(n)\}$. We will use P_k to stand for this function.

Example 5

We now turn our attention to the generating function for the sequence $\{p(n)\}$ of unrestricted partition numbers. We derive this by generalizing the argument used in the proof of Theorem 5.1.

We consider the infinite product

$$(1 + x + x^2 + \ldots)(1 + x^2 + x^4 + \ldots) \ldots\ldots$$

which we can write as

$$\prod_{t=1}^{\infty} (1 + x^t + x^{2t} + \ldots)$$

or even as

$$\prod_{t=1}^{\infty} \sum_{s=0}^{\infty} x^{st}.$$

The terms in this infinite product are obtained by taking terms of positive degree from a *finite* number of the infinite sums, in all possible ways, and multiplying these terms together. Thus the terms of degree n have the form

$$x^{m_1 \cdot 1} \cdot x^{m_2 \cdot 2} \ldots x^{m_k \cdot k}, \qquad \text{with } k, m_1, \ldots, m_k \in \mathbb{N}$$

where

$$m_1 \cdot 1 + m_2 \cdot 2 + \ldots + m_k \cdot k = n. \tag{5.8}$$

So, as before, the number of terms of degree n equals the number of solutions of equation (5.8), and hence is equal to the number of partitions of n. The only difference is that here k can also take any value from the set $\mathbb{N}^+$, and hence the number of solutions of equation (5.8) equals the number of *unrestricted* partitions of n, that is, $p(n)$.

Thus the generating function for the sequence $\{p(n)\}$ is

$$x \mapsto \prod_{t=1}^{\infty} (1 + x^t + x^{2t} + \ldots),$$

which we can also write as

$$x \mapsto \prod_{t=1}^{\infty} \frac{1}{(1 - x^t)}$$

or as

$$x \mapsto \frac{1}{(1 - x)(1 - x^2)(1 - x^3) \ldots}.$$

We use P to denote this function. The relationship between P and the generating functions, P_k, for the sequences, $\{p_k(n)\}$ of restricted partition numbers, is that, for $|x| < 1$,

$$P(x) = \lim_{k \to \infty} P_k(x).$$

This raises the question as to whether this limit actually exists. We could quote another theorem of analysis which does indeed guarantee that the limit does exist but, instead, we give a direct proof as it is not at all difficult.

For $0 < x < 1$, the terms $1/(1 - x^t)$ which occur in the product which defines $P_k(x)$ are all greater than 1. Hence $\{P_k(x)\}$ is an increasing sequence. Therefore, to show that it has a limit all we need do is to prove that is has an upper bound. In doing so we make use of the inequality,

$$1 + a < e^a, \qquad \text{for } a > 0, \tag{5.9}$$

which follows from the series for e^x by neglecting all but the first two terms.

Suppose now that $m \in \mathbb{N}^+$ and that $0 < x < 1$. Then $x^m < x$, and hence

$$\begin{aligned} \frac{1}{1 - x^m} &= 1 + \frac{x^m}{1 - x^m} \\ &< 1 + \frac{x^m}{1 - x} \\ &< e^{(x^m/(1-x))} \end{aligned}$$

by inequality (5.9). It follows that

$$\begin{aligned} P_k(k) &< \prod_{m=1}^{k} e^{(x^m/(1-x))} \\ &= e^{(x+x^2+ \dots +x^k)/(1-x)} \\ &< e^{x/(1-x)^2} \end{aligned} \tag{5.10}$$

since $(x + x^2 + \dots + x^k) < x/(1 - x)$.

Inequality (5.10) gives an upper bound for the sequence $\{P_k(x)\}$ and hence it follows that this sequence has a limit. It follows that the series

$$\sum_{n=0}^{\infty} p(n)x^n$$

does converge for $0 < x < 1$. Since the interval of convergence of a power series is centred on 0, it follows that this series converges for $|x| < 1$.

Exercise

5.2.1. Find the generating functions for the sequences

(a) $\{n(n - 1)\}$;
(b) $\{n^2\}$;
(c) $\{n^3\}$.

5.3 APPLICATIONS TO PARTITION NUMBERS

We are now able to show how generating functions can be used to give more information about partition numbers. Using algebraic methods to prove combinatorial results is, you may feel, somewhat contrary to the main spirit of this book. However, there is no point in tying our hands behind our backs and refusing to use these powerful methods. In some cases alternative combinatorial proofs are not difficult to find, and we shall point this out from time to time. Indeed, our first application is to prove algebraically, a result for which we have already given a combinatorial proof.

It follows from our description of the generating function P_k for the sequence $\{p_k(n)\}$ that

$$P_k(x) = P_{k-1}(x) \times \frac{1}{1 - x^k}$$

and hence

$$(1 - x^k)P_k(x) = P_{k-1}(x). \tag{5.11}$$

Equating the coefficients of x^n on both sides of equation (5.11) we obtain

$$p_k(n) - p_k(n - k) = p_{k-1}(n),$$

thus giving an alternative proof of Theorem 3.4.

Our next application is a little more complicated. It provides an upper bound for the partition numbers $p(n)$.

Theorem 5.2
For all $n \in \mathbb{N}$,

$$p(n) < e^{(1+(\pi^2/6))\sqrt{n}}.$$

Proof
Suppose $0 < x < 1$. Then

$$P(x) = \prod_{k=1}^{\infty} \frac{1}{(1 - x^k)}$$

and hence

$$\ln(P(x)) = \sum_{k=1}^{\infty} \ln\left(\frac{1}{1 - x^k}\right)$$

$$= \sum_{k=1}^{\infty} \sum_{m=1}^{\infty} \frac{(x^k)^m}{m}, \tag{5.12}$$

on using the standard power series for $\ln(1/(1 - a))$ $(= -\ln(1 - a))$.

All the terms in the double sum of equation (5.12) are positive. Hence we can interchange the order of summation to give

$$\ln(P(x)) = \sum_{m=1}^{\infty} \sum_{k=1}^{\infty} \frac{(x^k)^m}{m}$$

$$= \sum_{m=1}^{\infty} \frac{1}{m} \sum_{k=1}^{\infty} (x^m)^k$$

$$= \sum_{m=1}^{\infty} \frac{1}{m} \left(\frac{x^m}{1 - x^m} \right). \tag{5.13}$$

Now because $0 < x < 1$,

$$\frac{1 - x^m}{1 - x} = 1 + x + x^2 + \ldots + x^{m-1} > mx^{m-1}$$

and hence

$$\frac{x^m}{1 - x^m} < \frac{1}{m} \left(\frac{x}{1 - x} \right).$$

Therefore it follows from equation (5.13) that

$$\ln(P(x)) < \left(\frac{x}{1 - x} \right) \sum_{m=1}^{\infty} \frac{1}{m^2}$$

$$= \left(\frac{x}{1 - x} \right) \frac{\pi^2}{6}, \tag{5.14}$$

using the well-known fact that the series $\Sigma_{m=1}^{\infty} 1/m^2$ converges with sum $\pi^2/6$. It follows from 5.14 that

$$P(x) < \mathrm{e}^{(x/(1-x))\pi^2/6}. \tag{5.15}$$

Each term in the series for $P(x)$ is positive. Hence $P(x)$ is larger than any single term in the series. That is, for each $n \in \mathbb{N}$,

$$p(n)x^n < P(x),$$

and hence

$$p(n) < \frac{1}{x^n} P(x). \tag{5.16}$$

By inequalities (5.15) and (5.16), for $0 < x < 1$,

$$p(n) < \frac{1}{x^n} \mathrm{e}^{(x/(1-x))\pi^2/6}. \tag{5.17}$$

We now complete the proof by putting

$$x = \frac{\sqrt{n}}{\sqrt{n} + 1},$$

with $n \in \mathbb{N}^+$. Note that for this value of x we do have $0 < x < 1$. Using the fact that $\{(1 + 1/\sqrt{n})^{\sqrt{n}}\}$ is an increasing sequence with limit e, we deduce that

$$\frac{1}{x^n} = \left(\frac{\sqrt{n}+1}{\sqrt{n}}\right)^n = \left(\left(1 + \frac{1}{\sqrt{n}}\right)^{\sqrt{n}}\right)^{\sqrt{n}}$$
$$\leqslant \mathrm{e}^{\sqrt{n}}. \tag{5.18}$$

Also, for this value of x,

$$\frac{x}{1-x} = \sqrt{n},$$

and hence it follows from inequality (5.17) that

$$p(n) < \mathrm{e}^{\sqrt{n}} \cdot \mathrm{e}^{\sqrt{n}(\pi^2/6)}$$
$$= \mathrm{e}^{(1+(\pi^2/6))\sqrt{n}}.$$

The upper bound given for $p(n)$ in Theorem 5.2 is very crude. For example, it gives

$$p(100) < 30\,677 \times 10^{11}$$

whereas, in fact,

$$p(100) = 190\,569\,292.$$

This is only to be expected since the inequality $p(n) < (1/x^n)P(x)$, used in the proof came from neglecting all but one of the terms in the series for $P(x)$. Also the proof only uses only elementary methods. However, Theorem 5.2 is good enough to show that p does not grow as fast as the exponential function.

In section 5.5 we describe the Hardy–Ramanujan Formula which approximates $p(n)$ very accurately.

We can view the generating function, Q_k, for the sequence $\{q_k(n)\}$ of restricted partition numbers, namely

$$x \mapsto (1 + x^2 + x^3 + \ldots)(1 + x^2 + x^4 + \ldots)$$
$$\ldots (1 + x^k + x^{2k} + \ldots),$$

as being derived from the generating function, P, for the unrestricted partition numbers, namely

$$x \mapsto \prod_{t=1}^{\infty} (1 + x^t + x^{2t} + \ldots)$$

by dropping the terms from this function in line with the restrictions that we put on partitions in order to obtain the restricted partition numbers,

$q_k(n)$. Recall that $q_k(n)$ counts the number of partitions when we impose the restriction that there are no parts of size larger than k. The terms in the bracket

$$(1 + x^t + x^{2t} + \dots)$$

for $t > k$ correspond to parts of size greater than k in a partition. So it is these terms that we need to delete from the formula for P to obtain the formula for Q_k.

We can extend this idea and use it to obtain very readily the generating functions for other sequences of restricted partition numbers, as we illustrate with several examples.

We let $q^{\neq}(n)$ be the number of partitions of n into *unequal* parts, that is, parts whose sizes are all different. For example, we have the following partitions of 8 into unequal parts:

$$8$$
$$7 + 1$$
$$6 + 2$$
$$5 + 3$$
$$5 + 2 + 1$$
$$4 + 3 + 1,$$

hence $q^{\neq}(8) = 6$.

Which terms do we have to drop from P to obtain the generating function for the sequence $\{q^{\neq}(n)\}$? In the bracket

$$(1 + x^t + x^{2t} + \dots)$$

the different multiples of t in the exponents correspond to the number of parts of size t in the partition. Thus $1 = x^0$ corresponds to having 0 parts of size t, x^t to having one part of size t, x^{2t} to having two parts of size t, and so on. If all the parts have to be unequal we cannot have more than one part of size t. Hence we need to drop from this bracket all the terms other than the first two. Hence we obtain the generating function $Q^{\neq}$ for the sequence $\{q^{\neq}(n)\}$ by taking just the first two terms $(1 + x^t)$ from this bracket. Hence $Q^{\neq}$ is the function

$$Q^{\neq} : x \mapsto (1 + x)(1 + x^2)(1 + x^3) \dots$$

which we can write as

$$Q^{\neq} : x \mapsto \prod_{t=1}^{\infty} (1 + x^t).$$

If we restrict the parts in the partition to being unequal and also to being odd numbers then we need to delete from this product all those terms

$$(1 + x^2)(1 + x^4) \ldots$$

which correspond to the choice of a part whose size is an even number. Thus if we let $o^{\neq}(n)$ be the number of partitions of n into unequal, odd parts, then the generating function for the sequence $\{o^{\neq}(n)\}$ is

$$x \mapsto (1 + x)(1 + x^3)(1 + x^5) \ldots$$

Likewise if $e^{\neq}(n)$ is the number of partitions of n into unequal, even parts, then the generating function for the sequence $\{e^{\neq}(n)\}$ is

$$x \mapsto (1 + x^2)(1 + x^4)(1 + x^6) \ldots.$$

Further examples of generating functions obtained in this way are the subject of some of the Exercises at the end of this section. From the solutions to these Exercises it will be seen that algebraic manipulation of the formulas for generating functions can yield combinatorial relationships between different sorts of partition numbers.

Exercises

5.3.1. Let $o(n)$ = the number of partitions of n into parts whose sizes are all odd numbers. Find the generating function for the sequence $\{o(n)\}$. Hence prove that for all $n \in \mathbb{N}^+$,

$$o(n) = q^{\neq}(n).$$

5.3.2. Let $q^{\nmid 3}(n)$ be the number of partitions of n into parts whose sizes are not divisible by 3, and let $q^{<3}(n)$ be the number of partitions of n in which there are not more than two parts of the same size. Find the generating functions for the sequences $\{q^{\nmid 3}(n)\}$ and $\{q^{<3}(n)\}$. Hence show that for all $n \in \mathbb{N}^+$,

$$q^{\nmid 3}(n) = q^{<3}(n).$$

5.4 EULER'S IDENTITY

Up till now we have used the algebra of generating functions to derive information about the combinatorics of partitions. In this section we turn things the other way round. We use combinatorial arguments about partitions to derive an algebraic identity. We shall only be scratching the surface of a large subject.

The identity we are going to derive was first proved by the great eighteenth-century Swiss mathematician, Leonhard Euler (1707–1783). It comes from analysing dot diagrams in the appropriate way.

We consider partitions of n into unequal parts all of odd size. We let $o^{\neq}(n)$ be the number of such partitions. Using the methods of the previous section, it easily seen that the generating function, $O^{\neq}$, for the sequence $\{o^{\neq}(n)\}$ is given by

$$O^{\neq}(x) = (1 + x)(1 + x^3)(1 + x^5) \ldots \qquad (5.19)$$

We now derive a second expression for this generating function by analysing dot diagrams. We illustrate the method by considering a particular example of a partition into unequal odd parts, namely,

$$11 + 7 + 5 + 1 = 24$$

We use a dot diagram where, instead of using rows of dots to represent the terms in the partition, we use L-shaped arrays of dots, as shown in the diagram below.

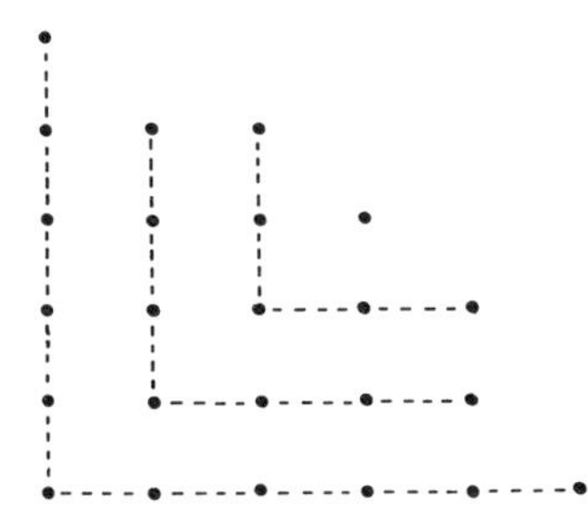

We can split up this diagram as follows. In the bottom left-hand corner there is a 4×4 array of dots. The dots to the right of the square, read in columns, form a partition of 4 into parts of size at most 4. The dots above the square, read in rows, form an identical partition. This is shown in the following diagram:

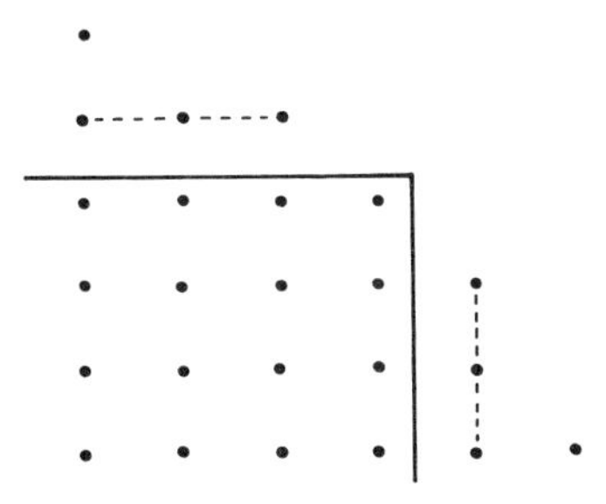

If we put these two identical partitions of 4, into parts of size at most 4 alongside one another, we get a partition of 8 into parts of size at most 8, as shown below.

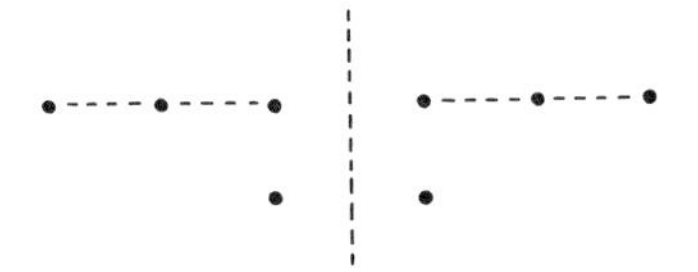

We now generalize this. Suppose we are given a partition of n into unequal odd parts. We can represent this partition by L-shaped arrays of dots. The bottom left-hand corner of this diagram will form a square array of dots. If k is the largest integer such that the diagram contains a $k \times k$ square array of dots, then clearly we must have $1 \leqslant k^2 \leqslant n$, and hence $1 \leqslant k \leqslant \text{int}(\sqrt{n})$. There will be $n - k^2$ dots not in this square. When the dots to the right of the square, taken in columns, are put together with the dots above the square, taken in rows, as we did above, we obtain a partition of $n - k^2$ into parts whose sizes are even and do not exceed $2k$.

In this way we obtain a one-to-one correspondence between partitions of n into unequal parts of odd size, and partitions of $n - k^2$ into even parts of size at most $2k$, for some k, $1 \leqslant k \leqslant \text{int}(\sqrt{n})$. Hence if $e^{\leqslant 2k}(n)$ is the number of partitions of n into even parts of size at most $2k$,

$$o^{\neq}(n) = \sum_{k=0}^{\text{int}(\sqrt{n})} e^{\leqslant 2k}(n - k^2). \tag{5.20}$$

(Note that we have started the sum at $k = 0$, rather than at $k = 1$, because we want equation (5.20) to hold not only when n is a positive integer, but also when $n = 0$. When $n \geqslant 1$, $e^{\leqslant 0}(n) = 0$, so including this term does not affect the sum in equation (5.20), and when $n = 0$, equation (5.20) becomes $o^{\neq}(0) = e^{\leqslant 0}(0)$, which holds by convention.)

From equation (5.20) it follows that

$$O^{\neq}(x) = \sum_{n=0}^{\infty} o^{\neq}(n)x^n = \sum_{n=0}^{\infty}\left(\sum_{k=0}^{\text{int}(\sqrt{n})} e^{\leqslant 2k}(n - k^2)\right)x^n. \tag{5.21}$$

We now change the order of summation on the right-hand side of equation (5.21). For a given value of k, the sum includes a term $e^{\leqslant 2k}(n - k^2)x^n$ provided that $k \leqslant \text{int}(\sqrt{n})$, that is, for $n \leqslant k^2$. Hence when we change the order of summation in equation (5.21) we get

$$\sum_{n=0}^{\infty} o^{\neq}(n)x^n = \sum_{k=0}^{\infty}\left(\sum_{n=k^2}^{\infty} e^{\leqslant 2k}(n - k^2)x^n\right). \tag{5.22}$$

Let $E^{\leqslant 2k}(x)$ be the generating function for the sequence $\{e^{\leqslant 2k}(n)\}$. Then

$$E^{\leqslant 2k}(x) = \sum_{n=0}^{\infty} e^{\leqslant 2k}(n)x^n$$

and hence

$$x^{k^2}E^{\leqslant 2k}(x) = \sum_{n=0}^{\infty} e^{\leqslant 2k}(n)x^{n+k^2} = \sum_{n=k^2}^{\infty} e^{\leqslant 2k}(n - k^2)x^n.$$

It therefore follows from equation (5.22) that

$$O^{\neq}(x) = \sum_{k=0}^{\infty} x^{k^2} E^{\leqslant 2k}(x).$$

Since, clearly,

$$E^{\leqslant 2k}(x) = \frac{1}{(1-x^2)(1-x^4)\ldots(1-x^{2k})}.$$

using formula (5.19) for $O^{\neq}(x)$ we have proved

Theorem 5.3 *Euler's Identity*

$$(1+x)(1+x^3)(1+x^5)\ldots$$
$$= 1 + \frac{x}{(1-x^2)} + \frac{x^4}{(1-x^2)(1-x^4)} + \frac{x^9}{(1-x^2)(1-x^4)(1-x^6)} + \ldots$$

5.5 THE HARDY–RAMANUJAN FORMULA

Most of the applications of generating functions in this chapter have used algebraic methods. The Hardy–Ramanujan asymptotic formula for $p(n)$, mentioned at the end of Chapter 4, arises from a sophisticated use of function theoretic methods. We give a brief sketch of their work, without going into any of the technical details.

So far we have thought of generating functions as having as their domains certain sets of real numbers. Since these functions are defined by power series, we can equally well regard them as complex functions, that is, as having sets of complex numbers as their domains. In this way the powerful methods of complex analysis become available. For example, the generating function for the sequence $\{p(n)\}$ of unrestricted partition numbers, regarded as a complex function, is the function

$$P : z \mapsto \prod_{k=1}^{\infty} \frac{1}{1-z^k}, \qquad \text{for } |z| < 1,$$

since the infinite product converges for all $z \in \mathbb{C}$, with $|z| < 1$. The coefficients in the power series $\Sigma_{n=0}^{\infty} a_n z^n$ for a complex function f are given by Cauchy's formula

$$a_n = \frac{1}{2\pi \mathrm{i}} \int_C \frac{f(z)}{z^{n+1}} \, \mathrm{d}z,$$

where C is a suitably chosen contour. Since $p(n)$ is the coefficient of z^n in the power series for the generating function P, it is given by

$$p(n) = \frac{1}{2\pi \mathrm{i}} \int_C \frac{P(z)}{z^{n+1}} \, \mathrm{d}z, \tag{5.23}$$

where C is any simple closed contour which surrounds the origin, and which is wholly enclosed inside the unit disc

$$D_1 = \{z \in \mathbb{C} : |z| < 1\}.$$

It follows from this that the values of $p(n)$ can be approximated by estimating the contour integral in equation (5.23). This is, however, more easily said than done. The biggest contributions to the values of this integral arise from parts of the contour nearest to the singularities of the integrand. It can be seen that the above function P has a singularity at each point $z \in \mathbb{C}$ where for some $k \in \mathbb{N}^+$, $z^k = 1$. Thus P has singularities at all the complex roots of 1, and these are densely distributed on the boundary of the unit disc, D_1. Thus every point on the boundary of D_1 is an essential singularity of P. This leads to tremendous technical difficulties.

In a celebrated paper, published in the *Proceedings of the London Mathematical Society*, in 1918, G. H. Hardy and S. Ramanujan* overcame these difficulties and derived a remarkable formula which approximates $p(n)$ very accurately. It is not possible to describe their method here. However it is possible to give some idea of the formula which they derived.

The first two terms of their formula are

$$\frac{1}{2\sqrt{2}\pi} \cdot \frac{\mathrm{d}}{\mathrm{d}n}\left(\frac{\mathrm{e}^{C\lambda_n}}{\lambda_n}\right) + \frac{(-1)^n}{2\pi} \cdot \frac{\mathrm{d}}{\mathrm{d}n}\left(\frac{\mathrm{e}^{C\lambda_n/2}}{\lambda_n}\right),$$

where $\lambda_n = \sqrt{n - (1/24)}$ and C is the constant $(\sqrt{2/3})\pi$.

If we carry out the differentiations indicated in this formula we obtain the formula

$$\frac{1}{4\pi\left(\sqrt{n - \frac{1}{24}}\right)^3}\left(\frac{1}{\sqrt{2}}\,\mathrm{e}^{\pi\sqrt{(2/3)[n-(1/24)]}}\left(\pi\sqrt{\frac{2}{3}\left(n - \frac{1}{24}\right)} - 1\right)\right.$$

$$\left. + (-1)^n\,\mathrm{e}^{(\pi/2)\sqrt{(2/3)[n-(1/24)]}}\left(\frac{\pi}{2}\sqrt{\frac{2}{3}\left(n - \frac{1}{24}\right)} - 1\right)\right) \qquad (5.24)$$

in all its gory detail. We will use $HR_2(n)$ to denote this formula.

From the first term of this formula Hardy and Ramanujan were able to derive their asymptotic formula for $p(n)$,

$$p(n) \sim \frac{1}{4\sqrt{3}}\left(\frac{\mathrm{e}^{\pi\sqrt{(2/3)n}}}{n}\right),$$

which we mentioned at the end of Chapter 4.

*G. H. Hardy and S. Ramanujan. 'Asymptotic Formulae in Combinatory Analysis', *Proceedings of the London Mathematical Society*, 2, 17, (1918), pp. 75–115.

The great accuracy of the Hardy–Ramanujan formula can be seen from the values in Table 5.1. The formula is so accurate that, for example,

$$p(50) = \text{int}\,(\text{HR}_2(50)).$$

In their paper Hardy and Ramanujan were able to prove that for every positive real number a, there is some positive integer N such that for all $n \geqslant N$,

$$p(n) = \text{int}\,(\text{HR}_k(n)),$$

where $k = \text{int}\,(a\sqrt{n})$ and HR_k denotes the sum of the first k terms of the Hardy–Ramanujan Formula.

The program given in the solution to Exercise 3.3.2 can be used to calculate the earlier values in this table. However it soon becomes hopelessly inefficient, since it involves counting the number of partitions of n by enumerating them all. More efficient algorithms make use of

Table 5.1 The Hardy–Ramanujan approximation for $p(n)$

n	$p(n)$	$\text{HR}_2(n)$
1	1	1.04038
2	2	2.06478
3	3	2.93888
4	5	5.02656
5	7	7.02937
6	11	10.93145
7	15	15.04265
8	22	22.05504
9	30	29.86557
10	42	42.06952
20	627	627.05738
30	5 604	5 603.54788
40	37 338	37 338.04032
50	204 226	204 226.7970
60	966 467	966 464.8362
70	4 087 968	4 087 969.043
80	15 796 476	15 796 477.76
90	56 634 173	56 634 167.93
100	190 569 292	190 569 293.7
200	3 972 999 029 388	3.97300×10^{12}
300	9 253 082 936 723 602	9.25308×10^{15}
400	6 727 090 051 741 041 926	6.72708×10^{18}

recurrence relations for the partition numbers. Using the recurrence relations given in Exercise 3.2.2 Hansraj Gupta† published tables of values of $p(n)$ for $1 \leqslant n \leqslant 600$. The later values in Table 5.1 are taken from Gupta's work.

5.6 THE STORY OF HARDY AND RAMANUJAN

The story of Hardy and Ramanujan has often been told, but it is well worth retelling.

Hardy was christened Godfrey Harold but he was almost universally known by his initials, G. H. He was born in Cranleigh, Surrey, on 7 February 1877. His father was art master and bursar at Cranleigh School. After attending this school, Hardy went as a scholar to Winchester, and then to Trinity College, Cambridge. Although his family was not wealthy, they had good connections in the English educational world, and after Hardy's mathematical talent had been spotted at an early age, there was no difficulty in obtaining for him the best mathematical education then available in England.

In fact, this was not at a very high level. Pure mathematics in Cambridge when Hardy arrived at Trinity College in 1896 was rather in the doldrums. Although England had produced some good mathematicians during the nineteenth century, they were cut off from continental European developments in mathematics, particularly developments in analysis associated with such mathematicians as Cauchy, Dedekind and Weierstrass, among others. The reason for this isolation goes back to the seventeenth century, when a dispute broke out between English and Continental mathematicians as to whether Newton or Leibniz should take the credit for having been first to 'invent' the calculus. To demonstrate their loyalty, English mathematicians continued to use Newton's notation, instead of the superior notation due to Leibniz (the $\mathrm{d}y/\mathrm{d}x$ notation which is still common today).

Hardy's introduction to modern mathematical analysis came from the applied mathematician, A. E. H. Love,‡ who advised him to read Jordan's *Course d'Analyse*. Hardy's first book, *A Course of Pure Mathematics*, first published in 1908, was one of earliest English texts to give a rigorous presentation of the basic concepts of mathematical analysis in the modern style. Unusually for a text at this level, Hardy's

†Hansraj Gupta, 'A Table of Partitions', *Proceedings of the London Mathematical Society*, 2, 39 (1935), pp. 142–9; and Gupta 'A Table of Partitions (II)', *Proceedings of the London Mathematical Society*, 2, 42, (1937), pp. 546–9.

‡Augustus Edward Hough Love was born in 1863. He became Sedleian Professor of Natural Philosophy in the University of Oxford in 1899 where he remained until his death in 1940. He is renowned especially for his work on the mathematical theory of elasticity.

book was still in print 80 years after it was first published.

Hardy remained as a lecturer in Cambridge until 1919, until he succeeded to the Savilian Chair of Geometry in Oxford. He returned to Cambridge in 1931 when he became the Sedleian Professor of Mathematics. He retired from this chair in 1942 and died in 1947.

Hardy's mathematical work is remarkable for two great collaborations. In 1912 he wrote the first of a long series of papers with J. E. Littlewood. Altogether Hardy and Littlewood published 93 joint papers (as well as two papers written together with G. Pólya, and one with E. Landau). Hardy, Littlewood and Pólya also jointly wrote the book *Inequalities*. Their collaboration continued for the rest of Hardy's life, their last joint paper being published in 1948 after Hardy's death. In the long history of mathematics there is no other example of such a successful partnership lasting so long.

Early in 1913, while Hardy was still in Cambridge, he received a letter from India. Thus began his association with the Indian mathematician, Srinivasa Ramanujan Aiyangar.

Ramanujan was born to a poor Brahmin family at Erode, a town 150 km south of Bangalore, in southern India, on 22 December 1887. His mathematical ability appeared early. After leaving school he entered the Government College at Kumbakonam in 1904 with a scholarship. Due to his neglect of non-mathematical subjects he failed an examination and lost his scholarship. For the same reason he failed another examination in 1907 and had to abandon the hope of further education, but he continued to work at mathematics. In 1909 he married and he needed to obtain a job to support his wife, Janaki. Eventually in 1912 he became a clerk with the Madras Port Trust.

In his search for employment in 1909, Ramanujan visited Mr V. Ramaswamy Ayar, the founder of the Indian Mathematical Society, and he subsequently also met Seshu Ayar. It was with their encouragment that on 16 January 1913 he wrote to Hardy enclosing over 100 mathematical results, taken from his notebooks.

Hardy recognized some of Ramanujan's formulas. Of others he wrote:

> The formulae (1.10)–(1.13) are on a different level and obviously both difficult and deep. An expert in elliptic functions can see at once that (1.13) is derived somehow from the theory of 'complex multiplication', but (1.10)–(1.12) defeated me completely; I had never seen anything in the least like them before. A single look at them is enough to show that they could only be written down by a mathematician of the highest class. They must be true because, if they were not true, no one would have had the imagination to invent them. Finally (you must remember that I knew nothing of Ramanujan, and had to think of every possibility), the writer must

> be completely honest, because great mathematicians are commoner than thieves or humbugs of such incredible skill.§

Hardy made strenuous efforts to bring Ramanujan to England, initially against Ramanujan's own wishes. However, by April 1914 Ramanujan was in Cambridge, with financial support from the Government of Madras and Trinity College. Although Ramanujan was now free from financial worries and could devote himself full-time to mathematics, his health soon deteriorated. He was elected a Fellow of the Royal Society and a Fellow of Trinity College in 1918. In 1919 his health had sufficiently recovered to enable him to travel home to India, but his health again failed and he died at a tragically early age on 26 April 1920 in Madras.

Hardy and Ramanujan are associated not only by their mathematical work, but also in one of the most often told stories about mathematicians. The story merits repetition, since, apart from anything else, it is marked out from many of the anecdotes which litter the history of mathematics by being true. We tell the story in Hardy's own words.

> I remember once going to see him when we was lying ill at Putney. I had ridden in taxi-cab No. 1729, and remarked that the number (7.13.19) seemed to me rather a dull one, and that I hoped it was not an unfavorable omen. 'No,' he replied, 'it is a very interesting number; it is the smallest number expressible as a sum of cubes in two different ways.'¶

Hardy summed up Ramanujan's mathematical achievement as follows:

> It was his insight into algebraical formulae, transformations of infinite series, and so forth, that was most amazing. On this side most certainly I have never met his equal, and I can compare him only with Euler or Jacobi. He worked, far more than the majority of modern mathematicians, by induction from numerical examples; all of his congruence properties of partitions, for example, were discovered in this way. But with his memory, his patience, and his power of calculation, he combined power of generalisation, a feeling for form, and a capacity for rapid modification of his hypotheses, that were often really startling, and made him, in his own peculiar field, without a rival in his day.
>
> It is often said that it is much more difficult now for a mathematician to be original than it was in the great days when the

§G. H. Hardy, *Ramanujan, Twelve Lectures on subjects suggested by his life and work*, Cambridge, 1940, p. 9.

¶$1729 = 1^3 + 12^3 = 9^3 + 10^3$. G. H. Hardy, P. V. Seshu Aiyar and B. M. Wilson (eds), *Collected Papers of Srinivasa Ramanujan*, Cambridge, 1927.

> foundations of modern analysis were laid; and no doubt in a measure it is true. Opinions may differ as to the importance of Ramanujan's work, the kind of standard by which it should be judged, and the influence which it is likely to have on the mathematics of the future. It has not the simplicity and the inevitableness of the very greatest work; it would be greater if it were less strange. One gift it has which no one can deny, profound and invincible originality. He would probably have been a greater mathematician if he had been caught and tamed a little in his youth; he would have discovered more that was new, and that, no doubt, of greater importance. On the other hand he would have been less of a Ramanujan, and more of a European professor, and the loss might have been greater than the gain.‖

Writing 13 years later, Hardy endorsed this judgement except for the last sentence which he described as 'ridiculous sentimentalism'.

‖G. H. Hardy, *Collected Papers of Srinivasa Ramanujan*, pp. xxxv–xxxvi.

6

Generating functions and recurrence relations

6.1 WHAT IS A RECURRENCE RELATION?

We have already met several examples of recurrence relations, as they arise naturally in many counting problems. In this chapter we look at recurrence relations more systematically. Our main concern will be with the use of generating functions to solve recurrence relations, but other methods are also described.

We begin with a typical problem which leads to a recurrence relation.

Problem 1

How many strings are there of n digits which do not contain consecutive zeros?

Solution

A string of n digits is simply a sequence $d_1 d_2 \ldots d_n$ where each d_i is one of the numbers 0, 1, 2, . . ., 9. We let a_n be the number of such strings which do not contain consecutive zeros. We calculate a_n by considering how strings of n digits, not containing consecutive zeros, can be built up from shorter strings with this property.

We can divide the strings of length n without consecutive zeros into two disjoint classes. The first class consists of all those strings which do not begin with a zero. The first digit can thus be any of the other nine and the remaining $n - 1$ digits must themselves form a string without consecutive zeros. There are a_{n-1} of these. Hence there are altogether $9a_{n-1}$ strings in this class.

The second class consists of those strings which start with a zero. The initial zero must be followed by one of the other 9 digits, and the remaining numbers make up a string of $n - 2$ digits not containing consecutive zeros. Thus there are $9a_{n-2}$ strings in this class.

It follows that for all $n \in \mathbb{N}^+$, with $n \geqslant 3$,

$$a_n = 9a_{n-1} + 9a_{n-2}. \tag{6.1}$$

This formula makes it quite straightforward to calculate the value of a_n for any particular value of n, given the additional facts that

$$\begin{aligned} a_1 &= 10 \\ a_2 &= 99, \end{aligned} \tag{6.2}$$

which can easily be checked. For example, we can calculate a_5 as follows:

$$\begin{aligned} a_3 &= 9a_2 + 9a_1 \\ &= 891 + 90 = 981. \\ a_4 &= 9a_3 + 9a_2 \\ &= 8829 + 891 = 9720. \\ a_5 &= 9a_4 + 9a_3 \\ &= 87\,480 + 8829 = 96\,309. \end{aligned}$$

Clearly this calculation could be extended as far as we like, but provides only a very long-winded way of calculating the values of a_n. In the subsequent sections of this chapter we discuss ways of converting equations (6.1) and (6.2) into a succinct formula which enables us to calculate the values of a_n very quickly. Before we do this, we want to use this example to help us decide what, in general, is meant by a recurrence relation for the terms of a sequence $\{a_n\}$.

We have seen that equation (6.1) enabled us to calculate a_3 from the values of a_1 and a_2 given by equations (6.2). We then used equation (6.1) again to calculate a_4 from a_2 and a_3, and again to calculate a_5 from a_3 and a_4. It is characteristic of a recurrence relation that it enables us to calculate each term in a sequence, $\{a_n\}$, from the values of the earlier terms in the sequence. As a first shot we could say that a recurrence relation has the form

$$a_n = f(a_{n-1}, a_{n-2}, \ldots) \tag{6.3}$$

where f is some given function. (In our equation (6.1), f is the function $f : (x, y) \mapsto 9x + 9y$). In order to make the definition as general as possible, our notation in equation (6.3) avoids specifying how many of the terms preceding a_n are needed to calculate the value of a_n. None the less equation (6.3) is not quite general enough. Consider the sequence of factorials, $\{n!\}$. This can be defined by the recurrence relation

$$a_n = na_{n-1} \tag{6.4}$$

together with the initial value $a_0 = 1$.

Notice that the definition of a_n in equation (6.4) involves, on the right-hand side, not only the preceding term in the sequence, a_{n-1}, but also the number n itself. So we need to revise equation (6.3) so as to include cases where the value of n enters into the formula for a_n. Thus a general recurrence relation has the form

$$a_n = f(n, a_{n-1}, a_{n-2}, \ldots) \tag{6.5}$$

where f is some given function. Again, our notation is deliberately vague about the number of terms of the sequence which are involved on the right-hand side of equation (6.5) as we wish to allow for such cases as

$$a_n = \sum_{k=1}^{n-1} a_k,$$

where the number of terms of the sequence which are involved in the definition of a_n can vary with the value of n.

The recurrence relation does not, by itself, determine a sequence. For example, equation (6.1) enables us to calculate a_3 from a_1 and a_2, and then a_4 from a_2 and a_3, and so on. But to get this process going we needed to be given the values of a_1 and a_2 to start with, as we were in equations (6.2). In general, a recurrence relation needs to be accompanied by one or more **initial conditions** which specify the first few terms of the sequence.

The recurrence relation with its initial conditions enables us to calculate all the terms of the sequence (assuming, of course, that we have a way to calculate the values of the function f which enters into the recurrence relation). By *solving* a recurrence relation we mean finding an explicit formula for a_n. (For what we mean here by an *explicit formula*, see the discussion in section 3.3.)

It is worth remarking here that what we have called a *recurrence relation* is, in other contexts, called a *recursive definition*. Recursive definitions are usually thought of as defining functions, with domain $\mathbb{N}$, rather than sequences, but, as we have already noted earlier, a sequence $\{a_n\}$ is really a function, $n \mapsto a_n$, in disguise. Recursive definitions are studied in the branch of mathematical logic known as *computability theory*, which is often called *recursive function theory*. Recursive definitions are allowed in several programming languages.

Recurrence relations are classified according to the form of the function f which occurs in the relation. In this chapter we discuss recurrence relations which can be solved by using the device of generating functions.

6.2 THE USE OF GENERATING FUNCTIONS

The basic idea of the generating function approach to recurrence relations is to translate the recurrence relation into an equation involving the generating function of the sequence. If we can extract from this equation an explicit formula for the generating function, we may be able to use this to derive a formula for the coefficients in its power series. These coefficients are, of course, just the terms of the sequence in which we are interested.

Before discussing this method in general we illustrate it in relation to the particular recurrence relation of section 6.1, as given by equation (1) and subject to the initial conditions (6.2).

We let A be the generating function for the sequence $\{a_n\}$. Thus

$$A(x) = \sum_{n=1}^{\infty} a_n x^n.$$

If we multiply both sides of the recurrence relation (6.1) by x^n, and sum for all integers $n \geqslant 3$, for which the relation (6.1) holds, we obtain

$$\sum_{n=3}^{\infty} a_n x^n = 9 \sum_{n=3}^{\infty} a_{n-1} x^n + 9 \sum_{n=3}^{\infty} a_{n-2} x^n. \tag{6.6}$$

The sum on the left-hand side of equation (6.6) is just the power series for A without the first two terms. Thus

$$\begin{aligned} \sum_{n=3}^{\infty} a_n x^n &= A(x) - a_1 x - a_2 x^2 \\ &= A(x) - 10x - 99x^2. \end{aligned}$$

To relate the term

$$9 \sum_{n=3}^{\infty} a_{n-1} x^n$$

which occurs on the right-hand side of equation (6.6) to the generating function A we need to pull out a factor x so that a_{n-1} multiplies x^{n-1}, as it does in the series for A. Thus

$$\begin{aligned} 9 \sum_{n=3}^{\infty} a_{n-1} x^n &= 9x \sum_{n=3}^{\infty} a_{n-1} x^{n-1} \\ &= 9x(A(x) - 10x). \end{aligned} \tag{6.7}$$

(Note that $\Sigma_{n=3}^{\infty} a_{n-1} x^{n-1} = a_2 x^2 + a_3 x^3 + \ldots$ and so is the power series for A without the first term, that is, $A(x) - a_1 x$, which is where we obtained equation (6.7) from.)

In a similar way,

$$9 \sum_{n=3}^{\infty} a_{n-2} x^n = 9x^2 \sum_{n=3}^{\infty} a_{n-2} x^{n-2} = 9x^2 A(x).$$

Thus we can deduce from equation (6.6) that

$$A(x) - 10x - 99x^2 = 9x(A(x) - 10x) + 9x^2 A(x). \tag{6.8}$$

It is now a straightforward matter to rearrange equation (6.8) to give

$$A(x) = \frac{1 + x}{1 - 9x - 9x^2} - 1. \tag{6.9}$$

Thus we have now achieved the first stage of our objective. We have obtained an explicit formula for the generating function of the sequence $\{a_n\}$. More than one method can be used to derive from this a formula for the coefficients in the corresponding power series. Probably the most efficient of these is to rewrite the formula for $A(x)$ using the technique of partial fractions.*

$1 - 9x - 9x^2 = (1 - \alpha x)(1 - \beta x)$, where α and β are the reciprocals of the solutions of the equation

$$1 - 9x - 9x^2 = 0. \tag{6.10}$$

Thus α and β are the solutions of the equation

$$y^2 - 9y - 9 = 0 \tag{6.11}$$

[We obtain equation (6.11) from (6.10) by putting $y = 1/x$.] It follows that $\alpha = \frac{1}{2}(9 + 3\sqrt{13})$, $\beta = \frac{1}{2}(9 - 3\sqrt{13})$. The standard partial fraction technique then yields

$$A(x) = \frac{1}{\alpha - \beta}\left(\frac{\alpha + 1}{1 - \alpha x} - \frac{\beta + 1}{1 - \beta x}\right) - 1 \tag{6.12}$$

From equation (6.12) we can very easily derive the power series expansion for A. It is

$$A(x) = \frac{1}{\alpha - \beta}\left((\alpha + 1)\sum_{n=0}^{\infty}(\alpha x)^n - (\beta + 1)\sum_{n=0}^{\infty}(\beta x)^n\right) - 1,$$

and, by equating coefficients we can deduce that, for $n \geqslant 1$,

$$a_n = \frac{1}{\alpha - \beta}\left((\alpha + 1)\alpha^n - (\beta + 1)\beta^n\right), \tag{6.13}$$

with α, β as given above. We have thus obtained an explicit formula for a_n.

The method that we have just used to solve the recurrence relation (6.1) will work for all recurrence relations of a similar type. In a moment we will be more precise about what this type is, but first it is worth commenting on a couple of features of the solution that we obtained in equation (6.13).

The recurrence relation (6.1), when taken with the initial conditions

*You will probably have used the method of partial fractions for evaluating the integrals of rational functions. If you have not met this method before, you will find it described in section 6.6.

(6.2), defines a sequence of positive integers. So it is somewhat surprising to find the irrational numbers α, β entering in to our formula for a_n. However, if in equation (6.13) we expand α^n and β^n in powers of $\sqrt{13}$, it will be seen that all the $\sqrt{13}$ terms cancel, and we are left with an expression in which $\sqrt{13}$ does not occur, albeit a rather more complicated expression that equation (6.13) itself. Another approach which also yields a solution in which there is no mention of irrational numbers is given in one of the Exercises at the end of this section.

We can use equation (6.13) to obtain very readily an asymptotic formula for a_n. We have (to three decimal places)

$$\beta = \tfrac{1}{2}(9 - 3\sqrt{13}) = -0.908,$$

and so $|\beta| < 1$. Thus

$$\lim_{n\to\infty} \beta^n = 0$$

and hence it follows from equation (6.13) that

$$a_n \sim \left(\frac{\alpha + 1}{\alpha - \beta}\right)\alpha^n,$$

that is

$$a_n \sim 1.008(9.908)^n,$$

using the numerical values for α and β, and working to three decimal places. The total number of sequences of n digits is 10^n. Thus we see that for small values of n, almost all sequences of n digits do not contain consecutive zeros, but, as we would expect, this proportion tends to 0 as the value of n increases.

Exercises

6.2.1. Write

$$\frac{1 + x}{1 - 9x - 9x^2}$$

as

$$(1 + x)\frac{1}{1 - 9x(1 + x)} = (1 + x)\sum_{n=0}^{\infty}(9x(1 + x))^n,$$

and, by equating coefficients, derive a formula for a_n.

6.2.2. *Fibonacci Numbers*

The Fibonacci numbers are one of the best-known sequences defined by a recurrence relation. Here is a counting problem in which they arise.

Rabbits take one month to reach maturity. A mature pair of rabbits produces another pair of rabbits each month. If you start with a pair of newly born rabbits, how many pairs will there be after n months?

Suppose that there are f_n pairs of rabbits after n months. The number of pairs after $n + 1$ months is the number, f_n, currently alive, plus the number born to those rabbits who have been alive for one month, namely, f_{n-1}. Hence we obtain the recurrence relation,

$$f_{n+1} = f_n + f_{n-1}.$$

We obtain the Fibonacci numbers when we solve this recurrence relation subject to the initial conditions

$$f_1 = 1$$

$$f_2 = 1.$$

(a) Find the generating function for the sequence $\{f_n\}$ of Fibonacci numbers.
(b) Hence find an explicit formula for f_n.
(c) Prove that the Fibonacci numbers are related to the binomial coefficients by the formula

$$f_{n+1} = \sum_{n/2 \leqslant k \leqslant n} C(k, n - k),$$

and interpret this formula in relation to Pascal's triangle.

6.3 HOMOGENEOUS LINEAR RECURRENCE RELATIONS

We noted in section 6.1 that the recurrence relation (6.1) is associated with the function

$$f : (x, y) \mapsto 9x + 9y.$$

This is a **linear function** with domain $\mathbb{R}^2$. In general, a linear function with domain $\mathbb{R}^k$ has the form

$$f : (x_1, \ldots, x_k) \mapsto \alpha_1 x_1 + \alpha_2 x_2 + \ldots + \alpha_k x_k,$$

where $\alpha_1, \ldots, \alpha_k$ are constants in $\mathbb{R}$. The recurrence relation corresponding to this function is

$$a_n = \alpha_1 a_{n-1} + \alpha_2 a_{n-2} + \ldots + \alpha_k a_{n-k} \qquad \text{for } n > k \qquad (6.14)$$

We will assume that here $\alpha_k \neq 0$ since otherwise equation (6.14) expresses a_n in terms of the preceding $k - 1$ terms of the sequence. We have written the equation in this form to bring out the fact that we can use it to calculate the value of a_n from the preceding k terms of the sequence, but in some ways it is more natural to rewrite this equation in a more symmetrical form, by not separating a_n from the other terms. This gives us an equation of the form

$$\alpha_0 a_n + \alpha_1 a_{n-1} + \alpha_2 a_{n-2} + \ldots + \alpha_k a_{n-k} = 0 \qquad (6.15)$$

where we shall also assume that $\alpha_0 \neq 0$ (so that the equation can be rearranged to give a_n in terms of $a_{n-1}, \ldots, a_{n-k}$). Equation (6.15) defines a linear recurrence relation of a particularly simple form. All the coefficients are constants and the right-hand side of the equation is 0. We call the relation expressed by equation (6.15) a **homogeneous linear recurrence relation with constant coefficients.** The word *homogeneous* is borrowed from the theory of differential equations and refers to the fact that the right-hand side of the equation is 0. For short, we will refer to such recurrence relations as *simple recurrence relations.*

Suppose we are given such a simple recurrence relation (6.15) subject to the initial conditions

$$a_1 = \beta_1, \ldots, a_k = \beta_k. \qquad (6.16)$$

We can easily solve this recurrence relation by means of the generating function method. We suppose that A is the generating function for the sequence $\{a_n\}$ defined by equations (6.15) and (6.16). Thus

$$A(x) = \sum_{k=1}^{\infty} a_n x^n.$$

If we multiply equation (6.15) by x^n and sum from $n = k + 1$ onwards, we obtain

$$\alpha_0 \sum_{n=k+1}^{\infty} a_n x^n + \alpha_1 x \sum_{n=k+1}^{\infty} a_{n-1} x^{n-1} + \alpha_2 x^2 \sum_{n=k+1}^{\infty} a_{n-2} x^{n-2} + \ldots + \alpha_k x^k \sum_{n=k+1}^{\infty} a_{n-k} x^{n-k} = 0$$

That is,

$$\begin{aligned} &\alpha_0(A(x) - \beta_1 x - \beta_2 x^2 - \ldots - \beta_k x^k) \\ &\quad + \alpha_1 x(A(x) - \beta_1 x - \beta_2 x^2 - \ldots - \beta_{k-1} x^{k-1}) \\ &\quad + \alpha_2 x^2(A(x) - \beta_1 x - \beta_2 x^2 - \ldots - \beta_{k-2} x^{k-2}) \\ &\quad + \ldots + \alpha_{k-1} x^{k-1}(A(x) - \beta_1 x) + \alpha_k x^k A(x) = 0 \end{aligned}$$

Rearranging the terms in this equation we get

$$\begin{aligned} &(\alpha_0 + \alpha_1 x + \ldots + \alpha_k x^k) A(x) \\ &\quad = \alpha_0\beta_1 x + (\alpha_0\beta_2 + \alpha_1\beta_1)x^2 + \ldots + (\alpha_0\beta_k + \ldots + \alpha_{k-1}\beta_1)x^k \end{aligned}$$

and thus

$$A(x) = \frac{\alpha_0\beta_1 x + (\alpha_0\beta_2 + \alpha_1\beta_1)x^2 + \ldots + (\alpha_0\beta_k + \ldots + \alpha_{k-1}\beta_1)x^k}{(\alpha_0 + \alpha_1 x + \ldots + \alpha_k x^k)}$$

Hence

$$A(x) = \frac{p(x)}{q(x)}$$

where

$$p(x) = \sum_{i=1}^{k} \left(\sum_{j=0}^{i-1} \alpha_j \beta_{i-j} \right) x^i$$

$$q(x) = \sum_{i=0}^{k} \alpha_i x^i.$$

We can then obtain a formula for a_n by calculating the coefficient of x in $A(x)$. We shall see later that the generating function method can be used to solve a wider class of recurrence relations. However, you may well think that it is rather a cumbersome method as it involves first calculating the generating function and only then a formula for a_n.

We can solve simple recurrence relations by a more direct method. We illustrate this method with the example with which we began this chapter, and we will then consider the theoretical justification for the method.

The recurrence relation we are considering can be written as

$$a_n - 9a_{n-1} - 9a_{n-2} = 0 \qquad \text{for } n \geqslant 3, \tag{6.17}$$

with the initial conditions given by equations (6.2).

We try to find a solution of equation (6.17) of the form

$$a_n = x^n \tag{6.18}$$

where x is some suitably chosen number. If this does indeed provide a solution of equation (6.17) we must have

$$x^n - 9x^{n-1} - 9x^{n-2} = 0.$$

Hence, assuming $x \neq 0$, by cancelling x^{n-2}, we get

$$x^2 - 9x - 9 = 0.$$

Note that, except for the change of y to x, this is the same as the quadratic equation (6.11) of section 6.2. So it has the same solutions

$$\alpha = \tfrac{1}{2}(9 + 3\sqrt{13}),$$

$$\beta = \tfrac{1}{2}(9 - 3\sqrt{13}).$$

Thus equation (6.17) has the two solutions

$$a_n = \alpha^n \quad \text{and} \quad a_n = \beta^n,$$

but we have yet to find a solution which is compatible with the initial conditions given by equation (6.2). We achieve this by combining the two solutions above, and we look for a solution of equation (6.17) of the form

$$a_n = A\alpha^n + B\beta^n, \tag{6.19}$$

where A, B are constants yet to be chosen so as to meet the initial conditions. (The theoretical justification for this approach will come later.) By putting $n = 1$ and $n = 2$ in equation (6.19), and using equations (6.2), we get

$$10 = A\alpha + B\beta$$

and

$$99 = A\alpha^2 + B\beta^2.$$

We can solve this pair of equations to obtain

$$A = \frac{99 - 10\beta}{\alpha(\alpha - \beta)}$$

$$B = \frac{99 - 10\alpha}{\beta(\beta - \alpha)}.$$

Hence the solution of the recurrence relation (6.17) which satisfies the initial conditions (6.2) is

$$a_n = \left(\frac{99 - 10\beta}{\alpha(\alpha - \beta)}\right)\alpha^n + \left(\frac{99 - 10\alpha}{\beta(\beta - \alpha)}\right)\beta^n, \tag{6.20}$$

At first sight this may look different to the solution (6.13) which we obtained in section 6.2. However, when we take account of the numerical values of α and β, it can readily be seen that these two solutions are identical.

If you have already met the 'try $y = e^{mx}$' method for solving linear differential equations with constant coefficients, that is, equations such as

$$\frac{d^2y}{dx^2} - 9\frac{dy}{dx} - 9y = 0,$$

you will notice a resemblance with the method we have just used to solve linear recurrence relations. Indeed, the two theories are almost exactly parallel, as they both concern linear operators on vector spaces. However it is not necessary to know about vector spaces in order to follow the piece of theory that is needed to justify the method we have just used. (See the Exercises at the end of this section for the vector space approach.)

Throughout the ensuing discussion we assume that we are dealing with the simple recurrence relation (6.15) subject to the initial conditions (6.16), above. The first theorem justifies our use of a solution of the form $a_n = x^n$.

Theorem 6.1
The sequence $\{x^n\}$ is a solution of equation (6.15) if x is a solution of the polynomial equation

$$\alpha_0 x^k + \alpha_1 x^{k-1} + \ldots + \alpha_{k-1} x + \alpha_k = 0. \tag{6.21}$$

Proof
Suppose that x is a solution of equation (6.21). Multiplying (6.21) by x^{n-k} we obtain

$$\alpha_0 x^n + \alpha_1 x^{n-1} + \ldots + \alpha_{k-1} x^{n-k+1} + \alpha_k x^{n-k} = 0.$$

Thus, if $a_n = x^n$ we do indeed have a solution of equation (6.15).

The next theorem justifies our decision to look for a solution of equation (6.15) of the form given by equation (6.19).

Theorem 6.2
If the sequences $\{x_n\}$ and $\{y_n\}$ are both solutions of equation (6.15), then for any choice of constants A, B, the sequence

$$\{Ax_n + By_n\} \tag{6.22}$$

is also a solution of equation (6.15).

Proof
If $\{x_n\}$ and $\{y_n\}$ are solutions of equation (6.15) then

$$\alpha_0 x_n + \alpha_1 x_{n-1} + \ldots + \alpha_k x_{n-k} = 0 \tag{6.23}$$

and

$$\alpha_0 y_n + \alpha_1 y_{n-1} + \ldots + \alpha_k y_{n-k} = 0 \tag{6.24}$$

Multiplying equation (6.23) by A, equation (6.24) by B and adding we obtain

$$A(\alpha_0 x_n + \alpha_1 x_{n-1} + \ldots + \alpha_k x_{n-k}) + B(\alpha_0 y_n + \alpha_1 y_{n-1} + \ldots + \alpha_k y_{n-k}) = 0,$$

that is

$$\alpha_0(Ax_n + By_n) + \alpha_1(Ax_{n-1} + By_{n-1}) + \ldots + \alpha_k(Ax_{n-k} + By_{n-k}) = 0,$$

which shows that the sequence (6.22) is indeed a solution of the recurrence relation (6.15).

In the language of vector spaces, we have shown that the solution set of equation (6.17) is **closed** under linear combinations, that is, it is a **linear subspace** of the vector space of all sequences $\{a_n\}$.

Theorems 6.1 and 6.2 show us how to obtain some of the solutions of the recurrence relation (6.15), but does this method yield *all* the solutions? Will this method enable us to find a solution compatible with whatever initial conditions are satisfied, and are there solutions which take a different form?

The easiest case to deal with is where the polynomial equation (6.21) of Theorem 6.1 has k distinct solutions. Note that our assumption that $\alpha_k \neq 0$ implies that these must be non-zero solutions. Suppose that these k distinct solutions of equation (6.21) are $x_1, x_2, \ldots, x_k$. Then for each choice of constant $A_1, A_2, \ldots, A_k$, the sequence

$$\{A_1x_1^n + A_2x_2^n + \ldots + A_kx_k^n\}$$

provides a solution of equation (6.15). It is then straightforward to check that, whatever initial conditions (6.16) are specified, it is always possible to find unique values of the constants $A_1, A_2, \ldots, A_k$ so that the sequence is a solution of the recurrence relation satisfying those particular initial conditions. In terms of vector spaces, this amounts to the fact that the solution space of equation (6.15) is k-dimensional and that the solutions $\{x_i^n\}$, for $1 \leqslant i \leqslant k$, form a basis for it (see the Exercises at the end of this section for the details). Notice that in this discussion we have not needed to assume that $x_1, \ldots, x_k$ are real numbers, so it also covers the case where some or all of the roots of the polynomial equation (6.21) are complex numbers.

The case where the polynomial equation (6.21) has multiple roots is only a little more complicated. It can be shown (again, see the Exercises for the details) that if x_0 is a root of the polynomial equation (6.21) of multiplicity r, then the sequences $\{x_0^n\}$, $\{nx_0^n\}$, $\ldots$, $\{n^{r-1}x_0^n\}$, are r linearly independent solutions of the recurrence relation (6.15).

Exercises

6.3.1. Find the generating function for the sequence $\{a_n\}$ given by

$$a_{n+2} = 3a_{n+1} + 4a_n$$

with $a_1 = 1$ and $a_2 = 3$. Hence find a formula for a_n.

6.3.2. Let a_n be the number of sequences of 0s, 1s and 2s of length n in which a 0 can only be followed by a 1. Find the recurrence relation which the sequence $\{a_n\}$ satisfies, and the generating function for the sequence.

6.3.3. Let A_n be the $n \times n$ matrix which has 1s on the leading diagonal, and on the diagonals immediately above and below the leading diagonal, and 0s everywhere else. Thus

$$A_n = \begin{bmatrix} 1 & 1 & 0 & 0 & 0 & . & 0 & 0 & 0 \\ 1 & 1 & 1 & 0 & 0 & . & 0 & 0 & 0 \\ 0 & 1 & 1 & 1 & 0 & . & 0 & 0 & 0 \\ 0 & 0 & 1 & 1 & 1 & . & 0 & 0 & 0 \\ . & . & . & . & . & . & . & . & . \\ . & . & . & . & . & . & . & . & . \\ 0 & 0 & 0 & 0 & 0 & . & 1 & 1 & 0 \\ 0 & 0 & 0 & 0 & 0 & . & 1 & 1 & 1 \\ 0 & 0 & 0 & 0 & 0 & . & 0 & 1 & 1 \end{bmatrix}.$$

Let $a_n = \det(A_n)$. Find a recurrence relation for the sequence $\{a_n\}$, and hence find a formula for a_n.

6.3.4. *Vector spaces and recurrence relations*
We let $\mathbb{R}^\infty$ be the set of all sequences, $\{a_n\}$, of real numbers. $\mathbb{R}^\infty$ becomes a *vector space* if we define vector addition by

$$\{a_n\} + \{b_n\} = \{a_n + b_n\}$$

and scalar multiplication by

$$\alpha\{a_n\} = \{\alpha a_n\}.$$

Theorem 6.2 shows that if we are given a simple recurrence relation

$$\alpha_0 a_n + \alpha_1 a_{n-1} + \alpha_2 a_{n-2} + \ldots + \alpha_k a_{n-k} = 0$$

then the subset, S, of $\mathbb{R}^\infty$ consisting of those sequences which satisfy this relation, is a subspace of $\mathbb{R}^\infty$.

(a) Prove that if the numbers $x_1, x_2, \ldots, x_t$ are all different, then the sequences $\{x_1^n\}, \{x_2^n\}, \ldots, \{x_t^n\}$ form a linearly independent subset of $\mathbb{R}^\infty$.

(b) For $1 \leqslant i \leqslant k$, let $\{y_n^i\}$ be the solution of the recurrence relation which satisfies the initial conditions

$$y_n^i = 0 \qquad \text{for } 1 \leqslant n \leqslant k \text{ and } n \neq i$$

and

$$y_i^i = 1.$$

Prove that the set of sequences

$$\{y_n^1\}, \{y_n^2\}, \ldots, \{y_n^k\}$$

forms a basis for the solution space S.

(c) Deduce that if $x_1, x_2, \ldots, x_k$ are k distinct solutions of the polynomial equation

$$\alpha_0 x^k + \alpha_1 x^{k-1} + \ldots + \alpha_{k-1} x + \alpha_k = 0,$$

then every solution of the recurrence relation has the form

$$\{A_1x_1^n + A_2x_2^n + \ldots + A_kx_k^n\}$$

for some choice of constants $A_1, \ldots, A_k$.

6.3.5. Prove that if x_0 is a non-zero root of the polynomial equation

$$\alpha_0x^k + \alpha_1x^{k-1} + \ldots + \alpha_{k-1}x + \alpha_k = 0$$

of multiplicity r, then the sequences

$$\{x_0^n\}, \{nx_0^n\}, \ldots, \{n^{r-1}x_0^n\}$$

provide r linearly independent solutions of the recurrence relation

$$\alpha_0a_n + \alpha_1a_{n-1} + \alpha_2a_{n-2} + \ldots + \alpha_ka_{n-k} = 0.$$

6.4 INHOMOGENEOUS LINEAR RECURRENCE RELATIONS

A linear recurrence relation with constant coefficients is called **inhomogeneous** if it has the form

$$\alpha_0a_n + \alpha_1a_{n-1} + \alpha_2a_{n-2} + \ldots + \alpha_ka_{n-k} = \beta(n) \qquad (6.25)$$

where β is some function, and $\alpha_1, \ldots, \alpha_k$ are constants. We allow the case where β is constant. As with the theory of linear differential equations, the solution of equation (6.25) is closely bound up with the solution of the associated homogeneous equation, obtained by replacing the $\beta(n)$ term by 0. The actual situation is described by the next theorem.

Theorem 6.3

Suppose that the sequence $\{c_n^0\}$ is a solution of equation (6.25). Then every solution of equation (6.25) has the form

$$\{b_n + c_n^0\}.$$

where $\{b_n\}$ is a solution of the homogeneous equation (6.15).

Proof

If $\{c_n^0\}$ is a solution of equation (6.25), then

$$\alpha_0c_n^0 + \alpha_1c_{n-1}^0 + \alpha_2c_{n-2}^0 + \ldots + \alpha_kc_{n-k}^0 = \beta(n). \qquad (6.26)$$

Now, if $\{b_n\}$ is a solution of the homogeneous equation (6.15),

$$\alpha_0b_n + \alpha_1b_{n-1} + \alpha_2b_{n-2} + \ldots + \alpha_kb_{n-k} = 0 \qquad (6.27)$$

Adding equations (6.26) and (6.27) gives

$$\alpha_0(b_n + c_n^0) + \alpha_1(b_{n-1} + c_{n-1}^0) + \ldots + \alpha_k(b_{n-k} + c_{n-k}^0) = \beta(n),$$

from which it follows that $\{b_n + c_n^0\}$ is a solution of equation (6.25).

Conversely, suppose that $\{d_n\}$ is a solution of equation (6.25). Thus

$$\alpha_0 d_n + \alpha_1 d_{n-1} + \alpha_2 d_{n-2} + \ldots + \alpha_k d_{n-k} = \beta(n). \qquad (6.28)$$

Substracting equation (6.27) from equation (6.28) gives

$$\alpha_0(d_n - c_n^0) + \alpha_1(d_{n-1} - c_{n-1}^0) + \ldots + \alpha_k(d_{n-k} - c_{n-k}^0) = 0,$$

which shows that if $b_n = d_n - c_n^0$, then $\{b_n\}$ is a solution of the homogeneous equation (6.15). Since $d_n = b_n + c_n^0$, this completes the proof that every solution of (6.25) is of this form.

Since we already know how to solve homogeneous recurrence relations, all that we need to be able to do to solve inhomogeneous equations is to find some way to come across a particular solution $\{c_n^0\}$. This can be done in several ways, and we illustrate several of these in connection with a particular example which arises from the following problem.

Problem 2

How many arithmetical operations are needed to evaluate the determinant of an $n \times n$ matrix, by row-reducing the matrix to upper triangular form, and then multiplying together the diagonal elements?

If you are not familiar with the linear algebra which lies behind this question, skip the solution to this problem, and start your reading again where we begin to discuss the solution of the recurrence relation (6.29) below which arises in this problem.

Solution

The first stage of the row-reduction process applied to the $n \times n$ matrix

$$\begin{bmatrix} a_{11} & \cdots\cdots & a_{1n} \\ \vdots & & \vdots \\ a_{n1} & \cdots\cdots & a_{nn} \end{bmatrix}$$

is to reduce it to the form

$$\begin{bmatrix} a_{11} & \cdots & \cdots & a_{1n} \\ 0 & a'_{22} & \cdots & a'_{2n} \\ \vdots & \vdots & & \vdots \\ 0 & a'_{n2} & \cdots & a'_{nn} \end{bmatrix}$$

by subtracting multiples of the top row from the remaining rows. (This assumes that $a_{11} \neq 0$. If $a_{11} = 0$, we need first to put a non-zero number in the first row and first column by interchanging the top row with the ith row, for some i such that $a_{1i} \neq 0$. Of course, if all the numbers in the first column are 0, the determinant of the matrix is 0, and we do not need to do any arithmetic to calculate it.)

This involves the following steps. First, for $2 \leqslant i \leqslant n$, we calculate the appropriate multiple of the first row that we need to subtract from the ith row. That is, we need to calculate

$$c_i = \frac{a_{i1}}{a_{11}}.$$

This involves $n - 1$ divisions. Then, for $2 \leqslant i \leqslant n$, we subtract c_i times the first row from the ith row. That is, we need to calculate, for $2 \leqslant i \leqslant n$ and $2 \leqslant j \leqslant n$,

$$a'_{ij} = a_{ij} - c_i a_{1j}.$$

This involves for each choice of i and j, one subtraction and one multiplication, and hence, altogether, $2(n - 1)^2$ arithmetical operations.

Having obtained the first row-reduced matrix in this way, we then need to calculate the determinant of the $(n - 1) \times (n - 1)$ matrix

$$\begin{bmatrix} a'_{22} & \cdots & a'_{n2} \\ \vdots & & \\ a'_{2n} & \cdots & a'_{nn} \end{bmatrix}$$

and then multiply this determinant by a_{11} to obtain the determinant of the original matrix. Hence, if we let r_n be the number of arithmetical operation which are needed to evaluate the determinant of this matrix, we see that

$$r_n = (n - 1) + 2(n - 1)^2 + r_{n-1} + 1,$$

that is

$$r_n - r_{n-1} = 2n^2 - 3n + 2, \qquad \text{for } n \geqslant 2. \tag{6.29}$$

Thus we have ended up with an inhomogeneous linear recurrence relation with constant coefficients. Note that the initial condition is

$$r_1 = 0, \tag{6.30}$$

as no arithmetic is needed to calculate the determinant of a 1×1 matrix.

Although equation (6.29) is a particularly simple example of an inhomogeneous recurrence relation, it will serve to illustrate different methods that are available for finding a particular solution.

6.4.1 Generating functions

We can use the generating function method equally well for inhomogeneous recurrence relations as for homogeneous ones. We let A be the generating function for the sequence $\{r_n\}$. From equation (6.29) we deduce that

$$\sum_{n=2}^{\infty} r_n x^n - x \sum_{n=2}^{\infty} r_{n-1} x^{n-1} = \sum_{n=2}^{\infty} (2n^2 - 3n + 2)x^n.$$

Thus, rewriting the left-hand side in terms of the generating function in the standard way (and remembering that $r_1 = 0$), we obtain

$$A(x) - xA(x) = \sum_{n=2}^{\infty} (2n^2 - 3n + 2)x^n$$

and hence

$$A(x) = \frac{1}{1-x}\left(\sum_{n=2}^{\infty} (2n^2 - 3n + 2)x^n\right). \tag{6.31}$$

From equation (6.31) we can proceed in two different directions, of which the second is generally to be preferred.

We can use our knowledge of some basic generating functions to rewrite the right-hand side of equation (6.31) as a rational function of x. Thus in section 5.2 we discovered that the generating functions for the sequences $\{n^2\}$, $\{n\}$, and $\{1\}$ are

$$x \mapsto \frac{x(1+x)}{(1-x)^3}, \quad x \mapsto \frac{\alpha}{(1-x)^2} \quad \text{and} \quad x \mapsto \frac{1}{1-x},$$

respectively. It thus follows that

$$\sum_{n=2}^{\infty} (2n^2 - 3n + 2)x^n = \frac{2x(1+x)}{(1-x)^3} - \frac{3x}{(1-x)^2} + \frac{2}{(1-x)} - x - 2$$

and we can then substitute this in equation (6.31) to obtain a formula for A as a rational function. However, this really leads us off in the wrong direction as our concern is not so much with the generating function as with its coefficients, and we can obtain a formula for these directly from equation (6.31) as follows.

In equation (6.31) we replace $1/(1-x)$ by its power series. At the same time it is convenient to change the summation variable from n to k. In this way we deduce from equation (6.31) that

$$A(x) = \left(\sum_{k=0}^{\infty} x^k\right)\left(\sum_{k=2}^{\infty} (2k^2 - 3k + 2)x^k\right). \tag{6.32}$$

We now equate coefficients in equation (6.32) to obtain

$$r_n = \sum_{k=2}^{n} (2k^2 - 3k + 2)$$

$$= 2 \sum_{k=2}^{n} k^2 - 3 \sum_{k=2}^{n} k + 2 \sum_{k=2}^{n} 1$$

$$= 2\left(\frac{n(n+1)(2n+1)}{6} - 1\right) - 3\left(\frac{n(n+1)}{2} - 1\right) + 2(n-1)$$

$$= \tfrac{2}{3}n^3 - \tfrac{1}{2}n^2 + \tfrac{5}{6}n - 1.$$

6.4.2 Spotting particular solutions

We have seen that the general solution of an inhomogeneous recurrence relation is made up of two parts: the general solution of the associated homogeneous recurrence relation, and a particular solution of the inhomogeneous relation. We have already described how to find the general solution of a homogeneous recurrence relation. In the current example it is particularly easy. The homogeneous recurrence relation associated with equation (6.29) is

$$r_n - r_{n-1} = 0,$$

which is particularly easy to solve! Its general solution is

$$r_n = A,$$

where A can be any constant. To complete the solution all we need to do is to find one particular solution of equation (6.29). How we do this does not matter; intelligent guesswork will do.

Since the right-hand side of equation (6.29) consists of a polynomial, it is reasonable to try to find a solution which is also a polynomial function, and as the polynomial in equation (6.29) is of degree 2, we try a polynomial of the same degree. Thus we aim to find a particular solution of equation (6.29) of the form

$$r_n = a + bn + cn^2,$$

for some constants a, b, c. Substituting this in equation (6.29) we see that a, b, c, need to be chosen so that

$$(a + bn + cn^2) - (a + b(n-1) + c(n+1)^2) = 2n^2 - 3n + 2$$

for all $n \geqslant 2$. That is

$$2cn + (b - c) = 2n^2 - 3n + 2, \qquad \text{for all } n \geqslant 2,$$

which is impossible, as we have a linear function on the left-hand side which cannot possibly be equal to the quadratic on the right-hand side for all $n \geqslant 2$. With a quadratic for r_n, $r_n - r_{n-1}$ turns out to be just a linear function. So it seems that to match the quadratic on the right-hand side, we need to take r_n to be a cubic. Thus we try

$$r_n = a + bn + cn^2 + dn^3.$$

This time, substituting in equation (6.29) we obtain

$$(a + bn + cn^2 + d^3) - (a + b(n - 1) + c(n - 1)^2 + d(n - 1)^3)$$
$$= 2n^2 - 3n + 2, \qquad \text{for all } n \geqslant 2.$$

That is,

$$3dn^2 + (2c - 3d)n + (b - c + d) = 2n^2 - 3n + 2, \qquad \text{for all n} \geqslant 2.$$

For the two polynomials to be equal for all integers $n \geqslant 2$, their coefficients must agree. Thus we need to have

$$3d = 2$$
$$2c - 3d = -3$$
$$b - c + d = 2,$$

from which it follows that $d = \frac{2}{3}$, $c = -\frac{1}{2}$ and $b = \frac{5}{6}$. The constant a can take any value, but this should not be surprising as the recurrence relation (6.29), without an initial condition specifying a value for a_1, does not define a unique sequence. Thus the general solution of equation (6.29) is

$$r_n = \tfrac{2}{3}n^3 - \tfrac{1}{2}n^2 + \tfrac{5}{6}n + A,$$

where A can be any constant. To satisfy the initial condition that $r_1 = 0$, we need to put $A = -1$. Thus the solution of equation (6.29) which satisfies the initial condition (6.30) is

$$r_n = \tfrac{2}{3}n^3 - \tfrac{1}{2}n^2 + \tfrac{5}{6}n - 1.$$

It is reassuring to see that this is the same as the solution which we obtained by the generating function method.

6.4.3 Difference equations

We have already noted the parallel between the theory of recurrence relations and that of differential equations. This parallel can be brought out even more sharply in the case of a recurrence relation such as equation (6.29) which involves the difference between consecutive terms in the sequence $\{r_n\}$. We can regard the operation of forming such differences as analogous to the operation of differentiation. If we use the symbol Δ to denote this operation, we can write Δr_n to stand for the difference $r_n - r_{n-1}$, and thus write the recurrence relation as

$$\Delta r_n = 2n^2 - 3n + 2. \qquad (6.33)$$

The theory of difference equations has been worked out in great detail. One of the pioneering works in this area is due to the mathematician George Boole,* who is more widely known for his contributions to mathematical logic. We will not go into this theory here, except to note that equation (6.33) is a difference equation of a particularly simple kind. Just as a differential equation such as

$$\frac{dy}{dx} = x^2$$

can be solved by direct integration, we can solve equation (6.33) by direct summation, as is shown by the following lemma.

Lemma 6.4
The solution of the difference equation

$$\Delta r_n = f(n), \qquad \text{for } n \geqslant 2,$$

subject to the initial condition

$$r_1 = f(1)$$

is

$$r_n = \sum_{k=1}^{n} f(k).$$

We omit the proof of this lemma as it is very straightforward. It can easily be checked that solving equation (6.29) by using the formula of the Lemma results in exactly the same summations as we carried out when we used the generating function method.

Exercises

6.4.1. Let a_n be the number of sequences of length n of letters of the English alphabet (which contains 26 different letters) in which between them the five vowels A, E, I, O, U occur an even number of times.
(a) Show that the sequence $\{a_n\}$ satisfies the recurrence relation

$$a_{n+1} = 16a_n + 5(26^n) \qquad \text{for } n \geqslant 1$$

with $a_1 = 21$.
(b) Find the generating function for the sequence $\{a_n\}$.
(c) Find a formula for a_n.

6.4.2. Suppose $m < n$. How many arithmetical operations are needed to row-reduce an $m \times n$ matrix to echelon form?

6.4.3. Let $\mathbf{A}$ be an invertible (i.e. non-singular) $n \times n$ matrix. The inverse of $\mathbf{A}$ may be calculated by forming the $n \times 2n$ matrix

$$(\mathbf{AI}_n)$$

*George Boole (1860) *A Treatise on the Calculus of Finite Differences*, Cambridge and London.

by juxtaposition of **A** with the $n \times n$ identity matrix, and row-reducing this matrix to echelon form, thus obtaining a matrix of the form

$$(\mathbf{I}_n \mathbf{B}),$$

where **B** is the inverse of **A**. How many arithmetic operations are needed to calculate the inverse of **A** by this method?

6.4.4. The Towers of Hanoi

Suppose that we have three pegs. On one of these pegs there are arranged n discs in order of size, with the largest disc at the bottom. The task is to transfer these n discs to the third peg, by moving one disc at a time, subject to the rule that no disc may ever be placed on top of a smaller disc.

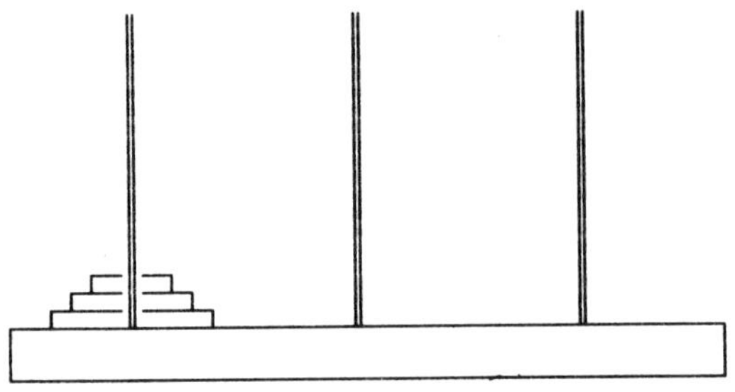

Let a_n be the minimum of moves needed to carry out this transfer. Find a recurrence relation for a_n. Hence, or otherwise, find a formula for a_n.

If, as the legend says, in the great temple of Benares the priests of Bramah are carrying out this task with 64 discs, and they move one disc each second, without making any mistakes, how long will it take before all the discs have been transferred and 'tower, temple and Brahmins alike will crumble into dust, and with a thunder clap the world will vanish'?

6.5 SOME NON-LINEAR RECURRENCE RELATIONS

When it comes to non-linear recurrence relations, life becomes much more difficult. In this section we consider non-linear recurrence relations of just one particular type. This type has been chosen because it does occur naturally in some counting problems, and, by good fortune, also succumbs to the generating function method.

We first need to explain the counting problem which gives rise to the recurrence relation we are going to study.

In a number of different mathematical contexts we use brackets to indicate the order in which certain operations are to be carried out, or how symbols are to be grouped. For example, in numerical work we meet such expressions as

$$(((5 \times 4) - 2) \div (2 + 7)),$$

in set algebra we have expressions such as

$$(((A \cap B) \cup C)\backslash(B \cap C)),$$

and in propositional logic we come across such formulas as

$$(((p \,\&\, q) \vee r) \rightarrow (q \,\&\, r)).$$

Although these three expressions come from different areas of mathematics, they share a common formal structure which corresponds to the sequence of brackets which is the same in each case, namely

$$((())()).$$

This is an example of a sequence of brackets that can occur in expressions of this sort. Other sequences such as

$$())(()$$

cannot occur in any legitimate expression. Here are some more examples of expressions of set algebra, giving us some more examples of legitimate bracket sequences.

$((((A \cap B) \cap C) \cap D) \cap E)$ with bracket sequence (((())))

$(((A \cap B) \cup (C \cap D)) \cap E)$ with bracket sequence ((()()))

$((A \cap B) \cup ((C \cap D) \cap E))$ with bracket sequence (()(()))

We call such sequences **legitimate bracket sequences**. Clearly in such a sequence the brackets occur in pairs, so that there are as many left-hand brackets, (, as there are right hand brackets,). We let b_n be the number of legitimate bracket sequences which contain n pairs of brackets. Our aim is to find an explicit formula for b_n.

We set about solving this problem by first deriving a recurrence relation for the sequence $\{b_n\}$. To do this we need to consider how expressions involving n pairs of brackets can be built up from shorter expressions. There is one point that we need to be careful about.

It is common mathematical practice to omit the outermost pair of brackets from an expression where this can be done without ambiguity. For example, it would be usual to write

$$(A \cap B) \cup (C \cap D)$$

rather than

$$((A \cap B) \cup (C \cap D)).$$

However, in defining legitimate bracket sequences, we assume that the rules for forming expressions require the outermost brackets in such a case. So ()() will not count as a legitimate bracket sequence. This stipulation implies that the number of pairs of brackets in an expression equals the number of operation symbols which occur in it. Thus an expression with n pairs of brackets, $n \geqslant 1$, has the form

$$(\$_1\$_2)$$

where $\$_1$, $\$_2$, are expressions which between them have $n-1$ pairs of brackets. So if $\$_1$ has k pairs of brackets, $\$_2$ will have $n-k-1$ pairs of brackets. Thus in $(\$_1\$_2)$, the bracket sequence in $\$_1$ can be chosen in b_k ways and that in $\$_2$, in b_{n-k-1} ways. Do all these choices lead to different bracket sequences in the resulting expression $(\$_1\$_2)$? It is easy to see that they do provided that both $\$_1$ and $\$_2$ contain at least one pair of brackets, that is, if $1 \leqslant k \leqslant n-2$. However, if $\$_1$ does not contain any brackets (for example, if it just consists of a set symbol, A, by itself) then the bracket sequence in $(\$_1\$_2)$ is exactly the same as in $(\$_2\$_1)$.

Thus, for $1 \leqslant k \leqslant n-2$, there are $b_k b_{n-k-1}$ different bracket sequences containing n pairs of brackets, of the form $(\$_1\$_2)$. In addition there are a further b_{n-1} sequences arising from the case where $\$_1$ does not contain any brackets, and $\$_2$ contains $n-1$ pairs of brackets. Thus we arrive at the recurrence relation

$$b_n = \sum_{k=1}^{n-2} b_k b_{n-k-1} + b_{n-1}, \qquad \text{for } n \geqslant 3. \tag{6.34}$$

Since there is just one legitimate bracket sequence with just one pair of brackets, (), and just one with two pairs of brackets, we have the initial conditions

$$\begin{aligned} b_1 &= 1 \\ b_2 &= 1. \end{aligned} \tag{6.35}$$

It can be seen that the recurrence relation (6.34) is highly non-linear. The number of terms on the right-hand side varies with n and these terms are quadratic rather than linear. None the less, it turns out that the generating function method can be used to solve this recurrence relation.

We let B be the generating function for the sequence $\{b_n\}$. Thus

$$B(x) = \sum_{n=1}^{\infty} b_n x^n.$$

We now adopt our standard technique of multiplying our recurrence relation (6.34) by x^n and summing for the appropriate range of values of n. In this way we obtain

$$\sum_{n=3}^{\infty} b_n x^n = \sum_{n=3}^{\infty} \left(\sum_{k=1}^{n-2} b_k b_{n-k-1} \right) x^n + \sum_{n=3}^{\infty} b_{n-1} x^n. \tag{6.36}$$

Taking account of the initial conditions (6.35), we see that

$$\sum_{n=3}^{\infty} b_n x^n = B(x) - x - x^2,$$

and that

$$\sum_{n=3}^{\infty} b_{n-1} x^n = x(B(x) - x),$$

but what about the remaining term with its complicated double summation?

At this stage we need to remember the formula for multiplication of power series that we discussed in section 5.1. We see from this that the coefficient of x^n involved in the remaining term of equation (6.36) is just the coefficient that arises when we multiply the power series $B(x)$ by itself. That is

$$(B(x))^2 = \sum_{n=2}^{\infty} \left(\sum_{k=1}^{n-1} b_k b_{n-k} \right) x^n$$

and hence

$$\sum_{n=3}^{\infty} \left(\sum_{k=1}^{n-2} b_k b_{n-k-1} \right) x^n = x(B(x))^2.$$

Therefore, from equation (6.36) we can deduce that

$$B(x) - x - x^2 = x(B(x))^2 + x(B(x) - x)$$

and hence that

$$x(B(x))^2 + (x - 1)B(x) + x = 0. \tag{6.37}$$

The outcome is that we have derived a quadratic equation for the values of the generating function. For each particular value of x ($\neq 0$) we can solve this equation by the standard formula to obtain

$$B(x) = \frac{1}{2x}\left((1 - x) \pm \sqrt{(1 + x)(1 - 3x)}\right) \tag{6.38}$$

The function B is defined by a power series and hence is continuous. To obtain a continuous function from equation (6.38) we need to make the same choice of sign, + or −, for each value of x. We know that the power series for B begins $x + x^2$, and from this we see that we need to take the minus sign in equation (6.38). Therefore

$$B(x) = \frac{1}{2x}\left((1 - x) - \sqrt{(1 + x)(1 - 3x)}\right) \tag{6.39}$$

We have obtained a comparatively simple formula for the generating function. But there is still quite a lot of work to be done before we can derive from it a formula for b_n. You will be pleased to see that we omit most of the gory details, and just give the bare outlines of the algebra.

It follows from the Binomial Theorem that for $|x| < 1$,

$$(1 - x)^{1/2} = 1 - \tfrac{1}{2}x - \sum_{n=2}^{\infty} \frac{(2n-3)!x^n}{n!(n-2)!2^{2n-2}}.$$

From this we can deduce that

$$\sqrt{(1+x)(1-3x)} = (1 - (2x + 3x^2))^{1/2}$$
$$= 1 - \tfrac{1}{2}(2x + 3x^2) - \sum_{n=2}^{\infty} \frac{(2n-3)!x^n(2+3x)^n}{n!(n-2)!2^{2n-2}},$$

and so

$$B(x) = x + \frac{1}{2x} \sum_{n=3}^{\infty} \frac{(2n-3)!x^n(2+3x)^n}{n!(n-2)!2^{2n-2}}$$

from which we can eventually deduce that

$$b_n = \left(\frac{3}{2}\right)^n \sum_{k=\operatorname{int}(n/2+1)}^{n+1} \frac{(2k-3)!}{(k-2)!(n+1-k)!(2k-n-1)!3^{k-1}}.$$

Exercise

6.5.1. Let the sequence $\{a_n\}$ be defined by

$$a_n = \sum_{k=1}^{n-1} a_k a_{n-k} \text{ for } n \geqslant 2,$$
$$a_1 = 1.$$

Find the generating function for this sequence. Hence find a constant α such that

$$a_n \sim \frac{\alpha 2^{2n}}{n^{3/2}}.$$

6.6 PARTIAL FRACTIONS

By a **rational function** is meant a function of the form

$$x \mapsto \frac{p(x)}{q(x)},$$

where p, q are polynomial functions.

The technique of **partial fractions** is a method of simplifying the algebraic expression for a rational function. It is used a good deal when it comes to integrating rational functions. The method, which we illustrate by a series of increasingly complicated examples, can be summed up as follows:

1. By dividing $q(x)$ into $p(x)$, rewrite the expression as
$$a(x) + \frac{r(x)}{q(x)},$$
where the degree of $r(x)$ is lower than the degree of $q(x)$.
2. Completely factorize the polynomial $q(x)$ into linear and quadratic factors with real number coefficients. (The general theory tells us that every polynomial with real coefficients can be factorized in this way; in practice, this may not be all that easy.)
3. Split $r(x)/q(x)$ into separate terms according to the factorization of $q(x)$. (What this means is explained in the following examples.)

Example 1

Consider the expression

$$\frac{1}{x^2 + x - 6}.$$

Here we have no need to carry our Step 1 as the numerator is already of lower degree than the denominator. The denominator factorizes as

$$(x + 3)(x - 2).$$

We now carry out Step 3. The rule for dealing with linear factors in the denominator is as follows:

Rule for Linear Factors

For each linear factor $(ax + b)$ which occurs to the power k in the denominator, introduce the terms

$$\frac{A_1}{(ax + b)}, \frac{A_2}{(ax + b)^2}, \ldots, \frac{A_k}{(ax + b)^k}.$$

In the present example, $k = 1$ for each of the linear terms in the denominator and so we split up the expression $1/(x^2 + x - 6)$ as follows:

$$\frac{1}{x^2 + x - 6} \equiv \frac{A}{(x + 3)} + \frac{B}{(x - 2)}.$$

The symbol $\equiv$ indicates that the expressions are identical. If we multiply through by $x^2 + x - 6$, we obtain

$$1 \equiv A(x - 2) + B(x + 3).$$

Since this is an identity, we can find the values of A and B by using either or both of two principles.

(a) The expressions will be equal whichever value we assign to x.
(b) If we regard the identity as an equation between polynomials, the coefficients on the left-hand side are equal to those on the right-hand side.

Here it is most efficient to find A and B in turn by substituting values of x which make the coefficient of the other constant disappear. Thus, putting $x = 2$ gives

$$1 = 5B$$

and hence $B = \frac{1}{5}$. And putting $x = -3$, we obtain

$$1 = -5A,$$

from which it follows that $A = -\frac{1}{5}$. Thus it follows that

$$\frac{1}{x^2 - x - 6} = \frac{-\frac{1}{5}}{x + 3} + \frac{\frac{1}{5}}{x - 2}.$$

Example 2
Consider the expression

$$\frac{x^3}{x^3 - 3x + 2}.$$

Here we have to carry out Step 1 first and rewrite this expression as

$$1 + \frac{3x - 2}{x^3 - 3x + 2}.$$

The denominator $(x^3 - 3x + 2)$ factorizes as $(x - 1)^2(x + 2)$. Since the linear term $(x - 1)$ occurs to the power 2, we need to include terms in our partial fraction expression with denominators $(x - 1)$ and $(x - 1)^2$. Thus we seek constants A, B and C so that

$$\frac{3x - 2}{x^3 - 3x + 2} \equiv \frac{A}{(x - 1)} + \frac{B}{(x - 1)^2} + \frac{C}{(x + 2)}.$$

Multiplying through by $(x - 1)^2(x + 2)$, we obtain

$$(3x - 2) \equiv A(x - 1)(x + 2) + B(x + 2) + C(x - 1)^2.$$

Now putting successively, $x = 1$ and $x = -2$, we get

$$1 = 3B$$
$$-8 = 9C,$$

from which it follows that

$$B = \tfrac{1}{3}$$
$$C = -\tfrac{8}{9}.$$

We could obtain an equation which enables us to find the value of A by substituting a third value for x. For example, putting $x = 0$, gives

$$-2 = -2A + 2B + C,$$

but it is rather easier to equate coefficients of x^2. If we do this we obtain the equation

$$0 = A + C,$$

from which it follows at once that $A = -C = \frac{8}{9}$. Thus

$$\frac{3x - 2}{x^3 - 3x + 2} = \frac{\frac{8}{9}}{x - 1} + \frac{\frac{1}{3}}{(x - 1)^2} - \frac{\frac{8}{9}}{x + 2}.$$

The remaining case we have to deal with is that where there are quadratic factors in the denominator. The way we deal with this is very similar to the way we handle linear factors.

Rule for Quadratic Factors

For each quadratic factor $(ax^2 + bx + c)$ which occurs to the power k in the denominator, introduce the terms

$$\frac{A_1x + B_1}{(ax^2 + bx + c)}, \frac{A_2x + B_2}{(ax^2 + bx + c)^2}, \ldots, \frac{A_kx + B_k}{(ax^2 + bx + c)^k}.$$

Example 3

We consider the expression

$$\frac{x + 1}{(x^2 + 1)(x + 4)}.$$

The quadratic factor occurs to the power 1, so we seek constants A, B and C so that

$$\frac{x + 1}{(x^2 + 1)(x + 4)} \equiv \frac{Ax + B}{(x^2 + 1)} + \frac{C}{(x + 4)}.$$

Multiplying through by $(x^2 + 1)(x + 4)$ we get

$$x + 1 \equiv (Ax + B)(x + 4) + C(x^2 + 1).$$

Putting $x = -4$, it follows that

$$-3 = 17C$$

whence $C = -\frac{3}{17}$. To obtain equations for A and B we equate the coefficients of x^2 and x. This gives

$$0 = A + C$$

and

$$1 = 4A + B,$$

respectively. From these equations we can deduce that $A = \frac{3}{17}$ and $B = \frac{5}{17}$. Thus, taking out the common factor $\frac{1}{17}$, we have

$$\frac{x + 1}{(x^2 + 1)(x + 4)} \equiv \frac{1}{17}\left(\frac{3x + 5}{(x^2 + 1)} - \frac{3}{(x + 4)}\right).$$

Exercises

Express each of the following in partial fraction form:

6.6.1. $$\frac{1}{(x + 3)(x - 7)},$$

6.6.2. $$\frac{x^2 + 3}{(x + 1)^2(x - 5)},$$

6.6.3. $$\frac{x^4 + 1}{(x^2 + 1)(x^2 + 2)},$$

6.6.4. $$\frac{1}{(x^2 + 1)(x + 1)^2}.$$

7

Permutations and groups

7.1 PERMUTATIONS

In the second half of this book we turn our attention to counting problems where ideas from group theory come in useful. Although probably most readers will already have met some group theory before, we are not going to assume any prior knowledge. Instead all the ideas about groups will be introduced as they are needed. We begin with a simple example that introduces **permutations** and our notation for them.

Problem 1

How many riffle shuffles does it take to restore a pack of cards back to its original starting position? Is there another shuffle which is better from this point of view?

First we had better explain what is meant by a **riffle shuffle**. When performed perfectly, a riffle shuffle involves splitting the pack into two equal sections which are then interleaved. This can be done in two ways: either so that the top card remains on top (very useful for card tricks) or so that it moves into second place. We will call these the **over** and **under** riffle shuffle, respectively. The diagram illustrates how these work in the case of a pack of 10 cards.

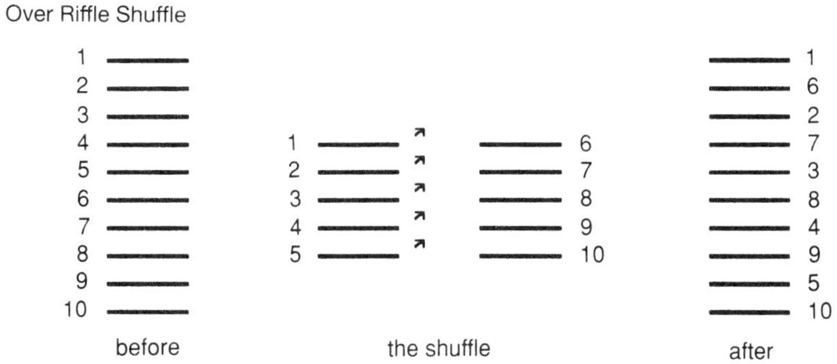

Over Riffle Shuffle

before	the shuffle		after
1			1
2			6
3			2
4	1	6	7
5	2	7	3
6	3	8	8
7	4	9	4
8	5	10	9
9			5
10			10

Under Riffle Shuffle

before	the shuffle		after
1	1	6	6
2	2	7	1
3	3	8	7
4	4	9	2
5	5	10	8
6			3
7			9
8			4
9			10
10			5

The effect of each of these shuffles is to rearrange the order of the cards. In Chapter 1 we called an arrangement in order of n objects, a *permutation* of those objects. It is now time to take a deeper look at permutations by considering them as functions.

The under riffle shuffle illustrated above replaces card 1 by card 6, card 2 by card 1, and so on. So the shuffle corresponds to the mapping:

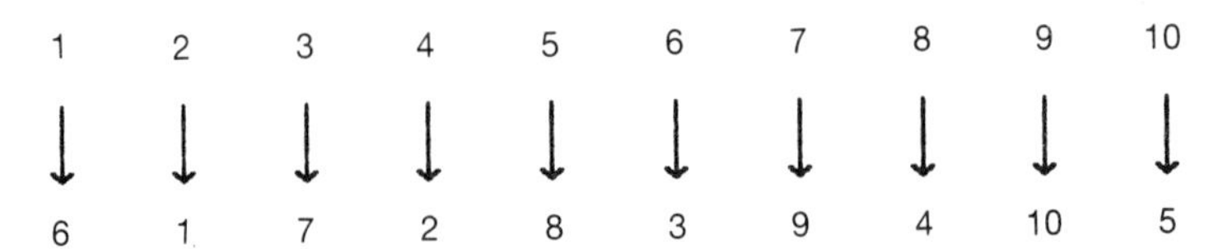

This mapping is a **bijection** (a one-to-one onto mapping) with domain and codomain the set $\{1, 2, 3, 4, 5, 6, 7, 8, 9, 10\}$. In general we define a **permutation** of a set X to be a bijection from X to X. We let $S(X)$ be the set of all permutations of the set X. We saw in Chapter 1 that if $\#(X) = n$, then $\#(S(X)) = n!$

We are mostly going to be interested in permutations in the cases where X is the set of the first n positive integers. In this case we use the notation S_n to stand for the set of all permutations of the numbers $\{1, 2, 3, \ldots, n\}$.

There are two different notations for permutations. The first is a minor variation of the arrow diagram above. We drop the arrows and put the numbers inside brackets. So the permutation corresponding to the under riffle shuffle of ten cards in this 'bracket notation' for permutations is

$$\begin{pmatrix} 1 & 2 & 3 & 4 & 5 & 6 & 7 & 8 & 9 & 10 \\ 6 & 1 & 7 & 2 & 8 & 3 & 9 & 4 & 10 & 5 \end{pmatrix}$$

However, this bracket notation for permutations is rather cumbersome. It is also not very helpful when it comes to answering questions about permutations. For example, the bracket notation does not make it very clear how many times we need to repeat a permutation to get back to the starting point. So we introduce a second much more useful way of representing permutations.

Since the domain and the codomain of a permutation are the same set, we can represent it in a single diagram with arrows showing how the elements of the set are mapped by the permutation. Consider, for example, the permutation which in bracket notation is

$$\sigma = \begin{pmatrix} 1 & 2 & 3 & 4 & 5 & 6 \\ 4 & 3 & 1 & 2 & 6 & 5 \end{pmatrix}.$$

We can represent the permutation σ by the diagram below

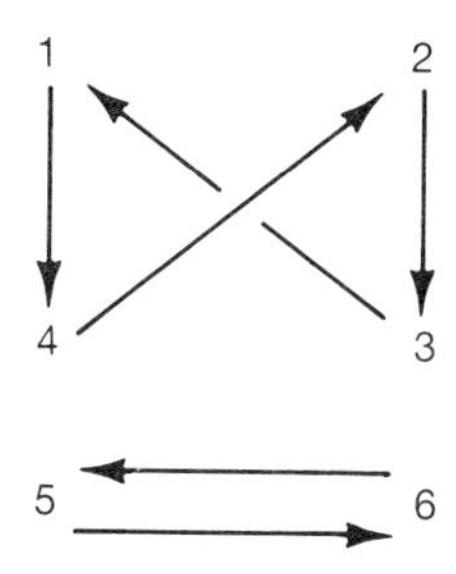

We see from this diagram that the effect of σ is to cycle the numbers $1, 4, 2, 3$ in this order, and to interchange the numbers 5 and 6. Thus the permutation is made up of two separate cycles. We represent this structure in the **cycle notation** in which we write the numbers in the two separate cycles in two separate brackets. Thus we write the permutation σ as

$$\sigma = (1\ 4\ 2\ 3)(5\ 6).$$

We call this the **disjoint cycle form** of σ (disjoint because the numbers making up the different cycles form disjoint sets).

This new notation has several advantages but we need to be careful with it. We are using a linear notation to represent cycles. Thus the bracket $(1\,4\,2\,3)$ represents the mapping $1 \rightarrow 4 \rightarrow 2 \rightarrow 3 \rightarrow 1$, and we need to remember that 3 is mapped back to 1 as this is not immediately obvious from the notation. Also, there is some arbitrariness in this notation. We did not have to write this cycle starting with 1. We could equally well have written it as $(3\,1\,4\,2)$, indicating the mapping $3 \rightarrow 1 \rightarrow 4 \rightarrow 2 \rightarrow 3$, which is just a different way of writing the same mapping. The same permutation could have been written in cycle notation in two other ways, since we could also write it as $(2\,3\,1\,4)$ or as $(4\,2\,3\,1)$. Thus there are four different ways of writing the same cycle of length 4. Similarly, the cycle of length 2 can be written in two different ways, as (5 6) or (6 5).

Also, we have a choice as to the order in which we write the two cycles involved in this permutation. We could have put the cycle (5 6) first, thus writing σ as (5 6)(1 3 2 4). So the same permutation can be written in several different ways in cycle notation. We will need to remember this when it comes to counting permutations.

There is another notational point to remember. Consider the permutation from S_6 whose bracket notation is

$$\begin{pmatrix} 1 & 2 & 3 & 4 & 5 & 6 \\ 2 & 3 & 1 & 5 & 4 & 6 \end{pmatrix}.$$

We can write this permutation in cycle notation as

$$(1\ 2\ 3)(4\ 5)(6).$$

The cycle (6) just containing the number 6 tells us that the permutation maps 6 to itself, i.e. that 6 is fixed by the permutation. It is usual to omit cycles of length 1 when writing permutations in cycle notation. So normally we would write the above permutation as

$$(1\ 2\ 3)(4\ 5).$$

Although it is convenient to omit cycles of length 1, it does introduce another possible ambiguity. At first sight, with the cycle (6) omitted, this looks like a permutation from S_5, as there is no mention of the number 6. So it is only really safe to omit cycles of length 1 if it is clear from the context which set of numbers it is that we are permuting.

After all this talk about the ambiguities in the cycle notation you may be wondering whether this second notation has any advantages at all. The obvious advantage is that it involves less writing than the bracket notation, but this is only a minor reason for preferring the cycle notation. We shall soon see that the cycle form of a permutation gives us a great deal of information about it. Indeed, the way a permutation splits up into cycles will be a continuing theme for the remainder of this book.

Exercises

7.1.1. Write each of the following permutations in cycle notation.

(a) $\begin{pmatrix} 1 & 2 & 3 & 4 & 5 & 6 & 7 & 8 & 9 & 10 \\ 3 & 1 & 8 & 9 & 7 & 4 & 10 & 2 & 6 & 5 \end{pmatrix}.$

(b) $\begin{pmatrix} 1 & 2 & 3 & 4 & 5 & 6 & 7 & 8 & 9 & 10 \\ 5 & 2 & 8 & 10 & 7 & 6 & 4 & 9 & 3 & 1 \end{pmatrix}.$

7.1.2. In how may different ways can the permutation of 1(a), above, be written in cycle notation?

7.1.3. If a permutation is chosen at random from S_n, what is the probability that it consists of a single cycle of length n?

7.2 GROUPS OF PERMUTATIONS

Since permutations are functions we can compose them in the usual way. Thus if σ and τ are permutations of the same set X we can define the composite permutation $\sigma \circ \tau$ by

$$\sigma \circ \tau(x) = \sigma(\tau(x)).$$

Permutations which are given to us in cycle form can be composed by tracing out what happens to each element in turn. When doing these calculations it is important to remember that the composite permutation $\sigma \circ \tau$ means *first* τ, *then* σ. Here is an example of how this works out in practice.

Let $\sigma = (1\,2\,4)(3\,5)$ and $\tau = (1\,4\,3\,5)$ be two permutations from S_5. We can calculate the composite permutation $\sigma \circ \tau$ as follows:

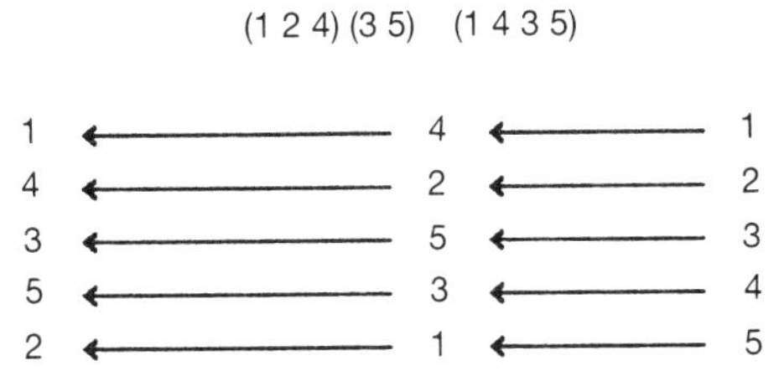

$$\begin{array}{ccccc} & (1\,2\,4)\,(3\,5) & & (1\,4\,3\,5) & \\ 1 & \longleftarrow & 4 & \longleftarrow & 1 \\ 4 & \longleftarrow & 2 & \longleftarrow & 2 \\ 3 & \longleftarrow & 5 & \longleftarrow & 3 \\ 5 & \longleftarrow & 3 & \longleftarrow & 4 \\ 2 & \longleftarrow & 1 & \longleftarrow & 5 \end{array}$$

We work from right to left as $\sigma \circ \tau$ involves first carrying out τ and then σ. The effect of τ on 1 is to map it to 4, and then σ maps 4 to 1, hence the combined effect is to map 1 to 1. That is

$$\sigma \circ \tau(1) = \sigma(\tau(1)) = \sigma(4) = 1.$$

We calculate the effect of $\sigma \circ \tau$ on each element of the set $\{1, 2, 3, 4, 5\}$ in turn, in a similar way. From the diagram above we can read off the bracket notation for $\sigma \circ \tau$ as

$$\begin{pmatrix} 1 & 2 & 3 & 4 & 5 \\ 1 & 4 & 3 & 5 & 2 \end{pmatrix}$$

which we can rewrite in cycle notation as $(2\,4\,5)$.

After a bit of practice it should not be necessary to write out the calculation of a composite permutation in full. The arrow diagram given above can be worked out mentally, and it should be possible to write down the composite permutation in cycle notation without the need to write down any intermediate steps.

Usually when writing composite permutations we omit the symbol $\circ$ and thus we write $\sigma\tau$ instead of $\sigma \circ \tau$. When the permutations are written in cycle notation this can lead to an ambiguity. For example if we write

$$(1\ 2\ 4)(3\ 5)(1\ 4\ 3\ 5),$$

this could mean

$$(1\ 2\ 4)(3\ 5) \circ (1\ 4\ 3\ 5)$$

or

$$(1\ 2\ 4) \circ (3\ 5)(1\ 4\ 3\ 5)$$

or even

$$((1\ 2\ 4) \circ (3\ 5)) \circ (1\ 4\ 3\ 5).$$

However, as we see from Lemma 7.4 below, composition of permutations satisfies the associativity property. Hence all these different expressions represent the same permutation, and so any ambiguity in the notation does not cause us any problems.

We have now seen that we can compose two permutations to obtain another permutation. This operation of **composition** of permutations has a number of important algebraic properties which we now list. It is important to know that these properties hold whichever set it is that we are permuting. Thus, we give full proofs of most of these properties.

Lemma 7.1

If $\sigma, \tau \in S(X)$ then $\sigma\tau \in S(X)$.

Proof

Suppose that $\sigma, \tau \in S(X)$. We need to show that $\sigma\tau$ is a bijection from X to X and hence is also in $S(X)$. We first show that $\sigma\tau$ is **injective** (one-to-one).

Let $x, y \in X$ with $x \neq y$. Then, as τ is injective $\tau(x) \neq \tau(y)$. Hence, as σ is injective $\sigma(\tau(x)) \neq \sigma(\tau(y))$. That is $\sigma\tau(x) \neq \sigma\tau(y)$. This shows that $\sigma\tau$ is injective.

Next we show that $\sigma\tau$ is **surjective** (onto). Suppose $z \in X$. Then, as σ is surjective, there is some $y \in X$ with $\sigma(y) = z$. Then, as τ is surjective, there is some $x \in X$ with $\tau(x) = y$. It follows that $\sigma\tau(x) = \sigma(\tau(x)) = \sigma(y) = z$. Hence z is the image under $\sigma\tau$ of an element from X. This shows that $\sigma\tau$ is surjective, and thus completes the proof that $\sigma\tau$ is a bijection and hence is in $S(X)$.

For each set X we let ι_X be the *identity* map from X to X. That is, it is the function defined by

$$\iota_X(x) = x \text{ for each } x \in X.$$

When it is clear from the context which set X is intended we will often write just ι instead of ι_X. In cycle notation, ι simply consists of cycles of length 1, and if we adopted the convention of not bothering to write these down, it would simply be represented by an empty space! This is not very convenient, so usually, when we are using the cycle notation, we represent ι by a single symbol, usually either ι or, often e.

It is easily seen that the following holds.

Lemma 7.2
$\iota_X \in S(X)$, and for all $\sigma \in S(X)$, $\sigma\iota_X = \sigma = \iota_X\sigma$.

Proof
Suppose $x, y \in X$ with $x \neq y$. Then $\iota_X(x) = x \neq y = \iota_X(y)$ and hence ι_X is injective. If $x \in X$, $\iota_X(x) = x$, and hence ι_X is surjective. Thus ι_X is a bijection from X to X, that is, $\iota_X \in S(X)$.

For each $\sigma \in S(X)$, and each $x \in X$, $\sigma\iota_X(x) = \sigma(\iota_X(x)) = \sigma(x)$, and hence $\sigma\iota_X = \sigma$. Similarly, $\iota_X\sigma = \sigma$.

Each bijection $\sigma \in S(X)$ has an inverse, σ^{-1}. The next Lemma is as straightforward as the previous one, and we therefore leave the proof to the reader.

Lemma 7.3
For each $\sigma \in S(X)$, σ^{-1} is also in $S(X)$, and $\sigma\sigma^{-1} = \iota_X = \sigma^{-1}\sigma$.

The last simple fact about the composition of permutations that we need to mention is that this operation has the associativity property. That is, for all permutations $\sigma, \tau, \rho \in S(X)$, we have

$$\sigma \circ (\tau \circ \rho) = (\sigma \circ \tau) \circ \rho.$$

In fact, as the proof shows, the associativity property holds in general for the composition of functions, whether or not the functions are permutations. (Of course, the domains and codomains of the functions must match up correctly for the composite functions to be defined.)

Lemma 7.4
For all $\sigma, \tau, \rho \in S(X)$, $\sigma \circ (\tau \circ \rho) = (\sigma \circ \tau) \circ \rho$.

Proof
Suppose $x \in X$. Then

$$\begin{aligned}(\sigma \circ (\tau \circ \rho))(x) &= \sigma((\tau \circ \rho)(x)) \\ &= \sigma(\tau(\rho(x))) \\ &= (\sigma \circ \tau)(\rho(x)) = ((\sigma \circ \tau) \circ \rho)(x),\end{aligned}$$

and hence $\sigma \circ (\tau \circ \rho) = (\sigma \circ \tau) \circ \rho$.

We have now shown that composition of permutations has the four properties given by Lemmas 7.1 to 7.4. These properties are so important that we give a special name to any collection of mathematical objects that can be combined in a way which satisfies them. We call such a structure a **group**. Thus we have the following definition.

A **group** is a pair $(G, *)$, where G is a set and $*$ is an operation defined on X which satisfies the following four properties.

G1 CLOSURE: For all $x, y \in G, x^*y \in G$.
G2 IDENTITY: There is some $e \in G$ such that for all $x \in G$ $x^*e = x$ and $e^*x = x$.

The special element e is called the **identity element** of G.
G3 INVERSES: For each $x \in G$ there is some $x^{-1} \in G$ such that $x^*x^{-1} = e$ and $x^{-1*}x = e$.

The element x^{-1} is called the **inverse** of x.
G4 ASSOCIATIVITY: For all $x, y, z \in G, x^*(y^*z) = (x^*y)^*z$.

The associativity property says that when it comes to combining three elements x, y, z, it does not matter how the expression is bracketed. Thus we can simply write

$$x^*y^*z$$

since it does not matter whether this is intended to mean $x^*(y^*z)$ or $(x^*y)^*z$. Similarly, we can omit brackets from expressions involving more than three group elements, such as $w^*x^*y^*z$.

We can sum up the four lemmas that we proved above by saying that

Theorem 7.5
For each set X, $(S(X), \circ)$ is a group, called the *group of permutations of X*.

Permutation groups form a very important collection of examples of groups. We shall meet some more examples of groups in the next chapter. Meanwhile, we give some further examples of groups, which occur in other branches of mathematics, though they will not be of great interest for us.

Example 1
$(\mathbb{Z}, +)$, the set of integers with the operation of addition, forms a group. Likewise $(\mathbb{Q}, +)$, the rational numbers, $(\mathbb{R}, +)$, the real numbers and $(\mathbb{C}, +)$, the complex numbers, all with the operation of addition, form groups. In each case the identity element is the number 0, and the inverse of x is $-x$.

Example 2
The sets which in Example 1 form groups with respect to addition, do not form groups with respect to the operation of multiplication. They are all closed under this operation. In each case the number 1 acts as an identity element, and multiplication, in each case, satisfies the associativity property. So the only thing that goes wrong is the existence of

inverses. In the case of the set $\mathbb{Z}$, this is an insurmountable problem. However, for $\mathbb{Q}$, $\mathbb{R}$ and $\mathbb{C}$ it is only the number 0 which does not have an inverse. Thus if we use $\mathbb{Q}^*$, $\mathbb{R}^*$ and $\mathbb{C}^*$, to denote the non-zero elements of these sets, then $(\mathbb{Q}^*, \times)$, $(\mathbb{R}^*, \times)$ and $(\mathbb{C}^*, \times)$ are three further examples of groups. It is from these examples, where the group operation is multiplication, that the notation x^{-1} comes from.

Example 3
For each positive integer n, $(\mathbf{M}_n, \times)$ is a group, where $\mathbf{M}_n$ is the set of $n \times n$ invertible (i.e. non-singular) matrices, and $\times$ is the usual operation of matrix multiplication. To be quite precise we should specify which field the matrix entries come from, but whichever choice we make, we get a group.

Example 4
For each positive integer n, $(\mathbb{Z}_n, +_n)$ is a group, where $\mathbb{Z}_n$ is the set $\{0, 1, 2, \ldots, n-1\}$ and $+_n$ is the operation of addition modulo n.

Example 5 *(Abstract groups)*
In Examples 1–4 the group elements were mathematical objects with which we were already familiar, and the operation was a natural one for those particular mathematical objects. Although, in a philosophical sense, mathematical objects are abstract objects, they seem very real to the mathematicians who use them. Thus groups of this kind are often referred to as **concrete groups**. These are contrasted with **abstract groups**, where we do not specify exactly what the elements of the group are, but only how they are combined. When the number of elements is small, this can conveniently be done by giving a table showing how the operation is defined. Here is an example of this kind.

The set $G = \{e, a, b, c, p, q, r, s\}$. The operation * is specified by the following table:

*	e	a	b	c	p	q	r	s
e	e	a	b	c	p	q	r	s
a	a	b	c	e	r	s	q	p
b	b	c	e	a	q	p	s	r
c	c	e	a	b	s	r	p	q
p	p	s	q	r	e	b	c	a
q	q	r	p	s	b	e	a	c
r	r	p	s	q	a	c	e	b
s	s	q	r	p	c	a	b	e

A table of this kind is called the multiplication table of the group or the **Cayley table** of the group (after the mathematician Arthur Cayley, a mathematician who lived between 1821 and 1895 and who did much pioneering work in abstract algebra, and especially with the theory of matrices). From the table we can see exactly how the operation $*$ is defined. The value of $x*y$ is found in the x-row and the y-column. Thus we see from this table that, for example, $q*c = r$.

Does the set form a group with respect to the operation $*$? The first three of the group properties are easy to check:

G1 Closure: We can see that all the entries in the table come from the set $\{e, a, b, c, p, q, r, s\}$, and hence, for all $x, y \in G$, $x*y \in G$.
G2 Identity: It is easy to see from the first row and column of the table that the element e of G acts as the identity element. Of course, this was made particularly easy to spot by the use of the standard symbol e for the identity element, and by placing it first in the table. But even if some other element had been the identity, it would not have been difficult to spot this from the table.
G3 Inverses: The identity element occurs once in each row and column of the table. Furthermore these occurrences of e are symmetrical about the diagonal which runs from the top left corner to the bottom right corner of the table. This shows that every element of G has an inverse in G.

In fact, every element of the group occurs once in each row and column of the table. (That this is always so for the multiplication table of a group follows from the result of Exercise 3 at the end of this section.) Thus the multiplication table of a group forms what is known as a **Latin square**.

To complete the proof that this table does indeed define a group we would need to check that the operation $*$ is associative. Unfortunately, there is no easy way to do this from the table. To check the associativity property we would need to check that for all choices of x, y, z from G, $x*(y*z) = (x*y)*z$. Since this includes those cases where two or three of x, y, z are equal, and there are 8 choices for each of x, y, z, there are $8^3 = 512$ equations to be checked! Some work can be saved by noting that in those cases where one or more of x, y, z is the identity element, e, the associativity property follows from the other group axioms. However, this still leaves $7^3 = 343$ equations to be checked, and as each equation involves four occurrences of $*$, we have to carry out the operation $*$ 1372 times. Clearly, this is not really a practical proposition. So you are asked to take it on trust for the time being that the table does indeed define a group (of course, you can check this if you wish!). We prove that it is indeed a group in the next section.

This difficulty arose in the case of a comparatively small group containing just eight elements. So for larger groups the situation would be even worse. Fortunately, in the case of concrete groups it is usually not difficult to see that the operation is associative. We have already noted that this follows automatically whenever the operation is composition of functions. In the case of abstract groups, rather than check all the possible cases, we look for short cuts.

Some more important examples of groups are described in the next section.

Exercises

7.2.1. Write each of the following permutations in disjoint cycle form.

(a) $(1\,2\,3\,4)(5\,6\,7)(1\,6\,7\,2\,9)(3\,4)$.

(b) $(1\,4\,9\,8\,6)(2\,3)(1\,5\,6)(7\,1\,3\,2)$.

7.2.2. We referred above to *the* identity element of a group, and to *the* inverse of an element x. Justify this terminology by proving that if $(G, *)$ is a group then the following must hold:

(a) The identity element is unique. That is, if for all $x \in G$, both

$$x*e_1 = x = e_1*x \text{ and } x*e_2 = x = e_2*x, \text{ then } e_1 = e_2.$$

(b) The inverse of each element is unique. That is, for each $x \in G$, if $y, z \in G$ are such that

$$x*y = e = y*x \text{ and } x*z = e = z*x, \text{ then } y = z.$$

7.2.3. Prove that if $(G, *)$ is a group then for all $x, y, z \in G$

(a) $x*y = x*z \Rightarrow y = z$,

(b) $y*x = z*x \Rightarrow y = z$.

(Note that it follows that in a group multiplication table, there are no repetitions in any row or column. So each row and each column consists of a permutation of the group elements. A theorem due to Cayley says that the correspondence between an element x of a group, and the permutation given by the x-row of the group table, is an isomorphism between the group and a group of permutations.)

7.2.4. A group $(G, *)$ is said to be **commutative** (or **Abelian**) if for all $x, y \in G$,

$$x*y = y*x.$$

Which of the groups mentioned in this section are commutative groups? (Note that because there are non-commutative groups when we specified the group axioms we needed to write the various equations both ways round. For example, we had to say that the inverse, x^{-1}, of a group element x, must satisfy both

$$x*x^{-1} = e \quad \text{and} \quad x^{-1}*x = e).$$

7.2.5. Give a multiplication table for an operation on the set $\{e, a, b, c, d\}$ which satisfies the three group axioms G1, G2 and G3, but which is not the Cayley table of a group because the associativity property is not satisfied.

7.3 SYMMETRY GROUPS

In this section we are going to describe an important class of groups. These are the groups which consist of all the symmetries of some geometrical figure (in either two or three dimensional space). We explain how these groups are defined by means of a typical example, the **symmetry group** of a square.

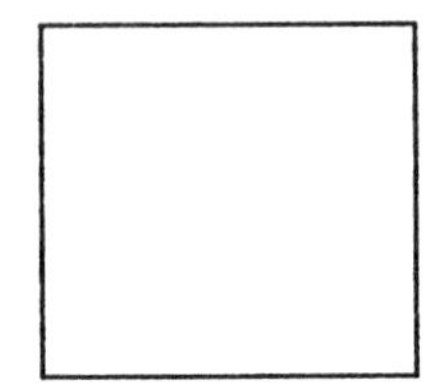

The square is a symmetrical figure, but what exactly do we mean by this? One way of explaining symmetry is to say that the square looks the same if we carry out certain geometrical transformations. For example, if we give the square a quarter turn (that is, if we rotate the square through an angle $\pi/2$, say, clockwise, about the axis through the centre of the square, and perpendicular to the plane of the square) it looks exactly the same as it did originally. All we need do is to turn this geometrical description of what a symmetry is into more technical mathematical language.

First we need to decide what we mean by a **geometrical transformation**. A geometrical transformation can move the figure around, but it must not change its size or shape, that is, it must not alter distances or angles. In fact, a transformation which leaves distances unaltered must also leave angles unchanged. So all we need to build into our definition of a geometrical transformation is that it leaves distances unchanged. Such transformations are called **isometries**.

If the transformation leaves the figure looking the same, the image of the transformation must be the same as the original set.

We are now ready to give the precise definition of what is meant by a **symmetry** of a figure $\mathcal{F}$ (where $\mathcal{F} \subseteq \mathbb{R}^2$ or $\mathcal{F} \subseteq \mathbb{R}^3$).

Definition

A symmetry of $\mathcal{F}$ is a mapping f with domain $\mathcal{F}$ such that

(a) f is an isometry, that is, for all $p, q \in \mathcal{F}$, the distance of $f(p)$ from $f(q)$ is the same as the distance of p from q.

(b) f is surjective, that is, $f(\mathcal{F}) = \mathcal{F}$.

It should be noted that because we are regarding symmetries as mappings (or functions), two transformations of a figure which are physically different, but which have the same effect on all points of the figure, count as being the same symmetry. Thus, for example, though rotating a square through an angle $\pi/2$ clockwise is physically different from rotating through an angle $3\pi/2$ anti-clockwise, since these two different motions have the same effect on each point of the square, they count as being the same symmetry. (This is nothing more than the usual convention that if f, g are mappings with the same domain D, and for each $d \in D, f(d) = g(d)$, then $f = g$, even though f and g are described in different ways.)

Thus the square has eight symmetries, which we will denote by symbols, as follows:

e	the identity ($e(x) = x$, for each point x of the square)
a	rotation through $\pi/2$ clockwise
b	rotation through π clockwise
c	rotation through $3\pi/2$ clockwise
p	reflection in the horizontal axis
q	reflection in the vertical axis
r	reflection in the diagonal axis shown
s	reflection in the diagonal axis shown

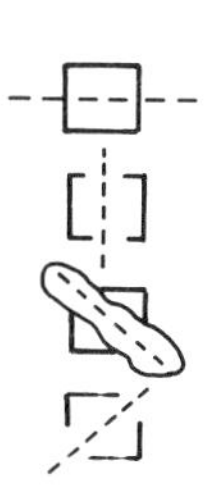

It is now quite straightforward to work out the Cayley table for the group of these symmetries (we have still to prove that they do form a group) by considering the effect of composing these symmetries. Don't forget that, as with our usual convention for composing functions, fg means *first do g, then do f*.

You should obtain precisely the same Cayley table as that of the abstract group given in the previous section. Symmetries are functions, and we know that composition of functions satisfies the associativity property. It follows that the operation defined by this Cayley table also has this property, and hence that it is indeed the Cayley table of a group.

We have thus seen that the symmetries of a square form a group. It is not difficult to prove that, in general, the symmetries of any figure form a group.

Theorem 7.6
Let $\mathcal{F}$ be some figure (i.e. $\mathcal{F} \subseteq \mathbb{R}^2$ or $\mathcal{F} \subseteq \mathbb{R}^3$). Then the set of all the symmetries of $\mathcal{F}$, with the operation of composition, form a group.

Proof
Let G be the set of all the symmetries of $\mathcal{F}$. We check that, with the operation of composition, G satisfies all the group axioms.

It is convenient to use the notation $d(p, q)$ for the distance between the points p and q. Now suppose $f, g \in G$. We show first that fg is an isometry. So take $p, q \in G$. Then we have

$$\begin{aligned} d(fg(p), fg(q)) &= d(f(g(p)), f(g(q)) \\ &= d(g(p), g(q)), \text{ as } f \text{ is an isometry,} \\ &= d(p, q), \text{ as } g \text{ is an isometry.} \end{aligned}$$

Hence fg is also an isometry. Also, $fg(\mathcal{F}) = f(g(\mathcal{F})) = f(\mathcal{F}) = \mathcal{F}$, and this completes the proof that fg is a symmetry of $\mathcal{F}$, that is, that $fg \in G$, and hence that the closure property is satisfied.

It is straightforward to check that the identity map $\iota_{\mathcal{F}}$ is a symmetry of $\mathcal{F}$, and hence that G has an identity element. We leave it as an exercise to check that if f is a symmetry of $\mathcal{F}$, then so also is f^{-1}, so that G satisfies the inverses property. We have already noted that the composition of symmetries is associative, and hence it follows that G is indeed a group.

It is worth noting that nothing in the above proof depended on the fact that we are dealing with figures in two-dimensional or three-dimensional space. It would work just as well with $\mathbb{R}^2$ or $\mathbb{R}^3$ replaced by $\mathbb{R}^n$, with $n \geqslant 4$. The only problem is that the geometry of figures in these higher-dimensional spaces is more difficult to picture!

If we introduce numerical labels we can easily relate symmetry groups to groups of permutations. For example, in the diagram of the square we can label the positions taken by the vertices with the numbers $1, 2, 3, 4$ as below.

Each symmetry of the square is specified as soon as we know the positions in which the vertices end up, when the symmetry has been carried out. Thus, for example, the symmetry a, corresponding to a rotation through $\pi/2$ clockwise, is fully described by saying that

the vertex at position 1 moves to position 2
the vertex at position 2 moves to position 3
the vertex at position 3 moves to position 4
the vertex at position 4 moves to position 1

(notice that we are regarding the numbers as labelling positions which are fixed in space).

Thus the symmetry a corresponds to the permutation in S_4 which can be written in cycle notation as (1 2 3 4). In a similar way each of the symmetries of the square corresponds to a permutation of the set $\{1, 2, 3, 4\}$. It is easy to see that this correspondence is as follows:

e	e
a	(1234)
b	(13)(24)
c	(1432)
p	(14)(23)
q	(12)(34)
r	(24)
s	(13)

It follows that these eight permutations by themselves form a group. They provide us with our first example of a **subgroup**, a concept which we describe in more detail in the next section.

Exercises

7.3.1. Prove that if f is a symmetry of the figure $\mathcal{F}$, then f has an inverse, f^{-1}, and f^{-1} is also a symmetry of $\mathcal{F}$.

7.3.2. How many symmetries are there in the symmetry group of a regular hexagon? Express each of these symmetries as permutations of the set $\{1, 2, 3, 4, 5, 6\}$. How many symmetries does a regular n-sided figure have?

7.3.3. Investigate the following groups:
(a) the symmetries of a tetrahedron;
(b) the rotational symmetries of a cube.

7.4 SUBGROUPS AND LAGRANGE'S THEOREM

Group theory is a very large subject with an enormous literature. I am going to confine myself to just those aspects of the theory which are relevant to the combinatorial problems that we are especially interested in. Before we can return to the problem about riffle shuffles which led us to groups in the first place, we need first to introduce a few more theoretical ideas about groups.

I have stressed that a group is a pair, $(G, *)$, consisting of a set G together with an operation $*$. However, whenever the operation can be understood from the context, it is convenient not to mention it explicitly. Thus we will often refer simply to 'the group G', leaving the operation $*$ to be understood. We have already seen that when composing permutations, we often omit the $\circ$ symbol. More generally, I will usually simply write xy rather than $x*y$.

Suppose, then, that G is a group. A subset H of G is called a **subgroup** of G, if H itself forms a group with respect to the same operation that makes G a group. (Note here the convenience of suppressing mention of the group operation. If we were being really pedantic, we would need to consider the group operation as a function $* : G \times G \to G$. Thus when it comes to H the relevant operation is the restriction of this function to the domain $H \times H$. So, in this terminology the subgroup is the structure $(H, * \upharpoonright (H \times H))$. Clearly, it is much easier simply to refer to the subgroup H.)

Thus H will be a subgroup of G if:

SG1. H is itself closed under the operation $*$, i.e. if for all $x, y \in H$, the element xy is also in H.
SG2. H contains the identity element e of G.
SG3. H contains inverses, i.e. for each $x \in H$, also $x^{-1} \in H$.

We do not need to specify that the elements of H satisfy the associativity property. G is a group and so whenever x, y and z are chosen from G, we have $x*(y*z) = (x*y)*z$. Hence whenever x, y and z are chosen from the subset H of G, they must also satisfy this equation. Thus the associativity property is passed down automatically from G to its subset H.

Note that H must be non-empty if it is to be a subgroup of G, since it must at least contain the identity element of G.

As is shown by Exercise 7.4.1, it is possible to replace the three subgroup conditions, SG1, SG2 and SG3 above, by a single condition.

Example 6

We have seen that $\mathbb{Z}$, $\mathbb{Q}$, $\mathbb{R}$ and $\mathbb{C}$ are all examples of groups, with the operation of addition in each case. Since $\mathbb{Z} \subseteq \mathbb{Q} \subseteq \mathbb{R} \subseteq \mathbb{C}$, it follows that $\mathbb{Z}$ is a subgroup of $\mathbb{Q}$, $\mathbb{R}$ and $\mathbb{C}$, that $\mathbb{Q}$ is a subgroup of $\mathbb{R}$ and $\mathbb{C}$, and that $\mathbb{R}$ is a subgroup of $\mathbb{C}$.

Example 7

Each group G with more than one element has two **trivial** subgroups. The set $\{e\}$ consisting of the identity element of G by itself can easily be seen to satisfy the subgroup conditions. Likewise the whole group G counts as being a subgroup of itself.

Example 8

The set of eight permutations from S_4 corresponding to the symmetries of a square, as listed in section 7.2 above, form a subgroup of S_4. The fact they form a group follows immediately from Theorem 7.6.

Example 9

Consider the group S_3 of all the six permutations of the set $\{1, 2, 3\}$. Rather than write out each of the permutations in S_3 in cycle notation, it is convenient to represent each permutation by a single symbol which we do as follows: $p = (123)$; $q = (321)$; $r = (12)$; $s = (23)$; $t = (13)$; and e is the identity element as usual. The Cayley table for this group is as follows:

$\circ$	e	p	q	r	s	t
e	e	p	q	r	s	t
p	p	q	e	t	r	s
q	q	e	p	s	t	r
r	r	s	t	e	p	q
s	s	t	r	q	e	p
t	t	r	s	p	q	e

It can be seen that the top left-hand corner of this table, taken by itself, is also the Cayley table of a group. In other words, the subset $\{e, p, q\}$ of S_3 forms a subgroup of S_3. In this case it was easy to recognize from the way that the Cayley table was set out, that this is a subgroup. In general it is rather difficult to spot subgroups from Cayley tables. Other ways of doing this will be described later.

In fact, apart from its two trivial subgroups, and the subgroup $\{e, p, q\}$ we have just found, the group S_3 has just three other subgroups, namely $\{e, r\}$, $\{e, s\}$ and $\{e, t\}$. Even for a small group like S_3, finding all the subgroups by brute force would be rather time-consuming. We need some theory to come to our aid, and here it is.

The group S_3 has six elements. Group theorists use their own term for the number of elements in a group. They call it the **order** of the group. Thus the order of a group G is just the number of elements in the set G. We do not need to introduce any new notation for this as we are already using $\#(G)$ for the number of elements in the set G.

What are the orders of the subgroups of S_3? The trivial subgroups $\{e\}$ and S_3 have orders 1 and 6 respectively. The other subgroups that we have found all have order either 2 or 3. There are no subgroups of order 4 or order 5. It turns out that there is a connection between the order of a group and the orders of its subgroups. You may feel bold enough to conjecture what this connection is on the basis of this single example. If not, you will discover what this connection is from the theory which follows.

Let G be a group and let H be a subgroup of G. (To make this abstract theory more concrete, it is suggested that you keep in mind a particular group, and a particular subgroup. We will use the example we have just described after giving the general definition.) For each $g \in G$, the **coset** gH is defined to be the set

$$\{gx \colon x \in H\}.$$

Thus gH is the set of all those elements of G obtained by combining the fixed element g with the elements of H in turn.

Example 10

We let G be the group S_3 with elements e, p, q, r, s, t as described above and we let H be the subgroup $\{e, r\}$. Thus we get the following cosets of H.

$$eH = \{ee, er\} = \{e, r\}$$

$$pH = \{pe, pr\} = \{p, t\}$$

$$qH = \{qe, qr\} = \{q, s\}$$

$$rH = \{re, rr\} = \{r, e\}$$

$$sH = \{se, sr\} = \{s, q\}$$

$$tH = \{te, tr\} = \{t, p\}.$$

Notice that in some cases the cosets of different elements turn out to be the same. For example, the cosets pH and tH both consist of the set $\{p, t\}$. (Since we are dealing with sets, the order in which elements occur in our calculation does not matter.) In other cases the cosets are completely different. For example there is no element of the group which is in both the cosets pH and qH; that is,

$$pH \cap qH = \varnothing.$$

We do not have two cosets which share some elements but which are not exactly the same. Also, each coset contains the same number of elements as the subgroup H, and each element of G occurs in at least one coset. Thus taking the distinct cosets, we obtain a partition of G, namely,

$$G = \{e, r\} \cup \{p, t\} \cup \{q, s\}$$

into three sets each containing $\#(H)$ elements, and hence

$$3 \times \#(H) = \#(G). \tag{7.1}$$

and from this it follows immediately that the number $\#(H)$ is a divisor of the number $\#(G)$. Of course, we do not need group theory to work out that 2 is a divisor of 6. The point of this observation is that the facts we have noticed about the cosets of the subgroup H are true in general, and we prove this below. Thus in each case we obtain an equation analogous to (7.1), with 3 replaced by some other integer, and hence we can make the same deduction about the relationship between the numbers $\#(G)$ and $\#(H)$.

It is possible to make sense of an equation such as (7.1) in the case where the groups G and H are infinite, but we shall confine our attention to the case of real interest where they are finite. Before reading the theory that follows, you may find it helpful to clarify your ideas by calculating the cosets of the subgroup $\{e, p, q\}$ of S_3, or by doing a similar calculation with some other subgroup of a finite group.

Since we are interested in cases where two cosets turn out to be identical, it is helpful to begin with a general criterion for this to be the case.

Lemma 7.7

If H is a subgroup of the group G, and $g_1, g_2 \in G$, then

$$g_1 H = g_2 H \Leftrightarrow g_2^{-1} g_1 \in H. \tag{7.2}$$

Proof

Suppose, first, that $g_1 H = g_2 H$. Since $e \in H$, $g_1 e = g_1 \in g_1 H$, and hence $g_1 \in g_2 H$. Thus for some $h \in H$, $g_1 = g_2 h$. It follows that $g_2^{-1} g_1 = h$, and hence that $g_2^{-1} g_1 \in H$. This proves one half of the double implication (7.2).

Now suppose that $g_2^{-1} g_1 \in H$, say, $g_2^{-1} g_1 = h$, so that $g_1 = g_2 h$. We show that it follows that $g_1 H = g_2 H$. Suppose $x \in g_1 H$. Then, for some $h_1 \in H$, $x = g_1 h_1$. Hence $x = (g_2 h) h_1 = g_2 (h h_1)$. Now $h h_1 \in H$, as H is a subgroup of G and therefore satisfies the closure condition. It follows that $x \in g_2 H$. This shows that $g_1 H \subseteq g_2 H$.

We can rewrite the equation $g_1 = g_2 h$ as $g_2 = g_1 h^{-1}$. H is a subgroup of G and hence h^{-1} is also in H. We can thus repeat the argument in the previous paragraph to show that also $g_2 H \subseteq g_1 H$. From $g_1 H \subseteq g_2 H$ and $g_2 H \subseteq g_1 H$ it follows that $g_1 H = g_2 H$, and this completes the proof.

We are now ready to prove that the facts about cosets which we noticed in the particular example above, are true in general.

Lemma 7.8

If H is a subgroup of the group G, then any two cosets of H are either identical or completely different. That is, for all g_1, $g_2 \in H$, either $g_1 H = g_2 H$ or $g_1 H \cap g_2 H = \varnothing$.

Proof

We show that if $g_1 H \cap g_2 H \neq \varnothing$, then $g_1 H = g_2 H$. Suppose that there is some $x \in g_1 H \cap g_2 H$. Then for some h_1, $h_2 \in H$, $x = g_1 h_1$ and $x = g_2 h_2$. Since $g_1 h_1 = g_2 h_2$, it follows that

$$g_2^{-1} g_1 = h_2 h_1^{-1}.$$

Because H is a subgroup of G, $h_2h_1^{-1} \in H$, and hence $g_2^{-1}g_1 \in H$. Therefore, by Lemma 7.6, $g_1H = g_2H$.

Lemma 7.9
If H is a subgroup of G, each element of G is in at least one coset of H.

Proof
As H is a subgroup of G, it contains the identity element e of H. Hence for each $g \in G$, $g = ge \in gH$.

Lemma 7.10
If H is a subgroup of G and $\#(H) = n$, then for each coset gH of H, $\#(gH) = n$.

Proof
Suppose $H = \{h_1, \ldots, h_n\}$, with the h_i all distinct. Then for $g \in G$, $gH = \{gh_1, \ldots, gh_n\}$. The elements gh_i are also all distinct, as if $gh_i = gh_j$, then $h_i = h_j$. Hence $\#(gH) = n$.

The number of distinct cosets of H in G is called the **index** of H in G, written $|G : H|$. It follows from the lemmas above that the distinct cosets of H partition G into $|G : H|$ sets each containing $\#(H)$ elements. Therefore, we have

$$|G : H| \times \#(H) = \#(G).$$

From this equation we can deduce the following important theorem.

Theorem 7.11 *(Lagrange's Theorem)*
If H is a subgroup of the finite group G, then $\#(H)$ is a divisor of $\#(G)$.

This theorem is often expressed as 'the order of a subgroup divides the order of the group'. Notice that Lagrange's Theorem only tells us about the *possible* orders of the subgroups of a given group G. It says that G can only have a subgroup of order n if n is a divisor of $\#(G)$. It does *not* say that if n is a divisor of $\#(G)$, then G must have a subgroup of this order. The smallest example which shows this is that of a group of order 12 which does not have a subgroup of order 6, even though 6 is a divisor of 12. This example can be found in Exercise 7.4.4 below.

Exercises

7.4.1. Prove that if G is a group and H is a non-empty subset of G, then H is a subgroup of G if and only if it satisfies the following condition:

$$\text{for all } g, h \in H, gh^{-1} \in H.$$

7.4.2. Let G be the abstract group whose Cayley table is given in section 7.2. Show that $\{e, b\}$ is a subgroup of G, and find all the distinct cosets of this subgroup.

7.4.3. Find as many subgroups of S_4 as you can. [You will need to make use of a certain amount of trial and error, but the more theory you know, the less work is needed. You may find it helpful to postpone tackling this question until after you have read section 7.5.]

7.4.4. Consider the following set, H, of 12 permutations from S_4: {e, (12)(34), (13)(24), (14)(23), (123), (132), (124), (142), (134), (143), (234), (243)}. H forms a subgroup of S_4. (This can be proved most directly by observing that these permutations correspond to the rotational symmetries of a regular tetrahedron.) Show that this group of order 12 does not have a subgroup of order 6.

7.5 ORDERS OF GROUP ELEMENTS

Let G be a group and let g be an element of G. Since G satisfies the closure condition, G also contains the element gg, and hence also ggg (remember that because of the associativity property, we can omit any brackets), $gggg$, and so on.

At this point it is convenient to use the notation g^n to stand for the result of combining n gs together. Thus g^2 stands for gg, g^3 for ggg, and so on. In addition, we use g^1 to stand for g. We call the elements of G of the form g^n, **powers** of g.

If G is a finite group, the elements $g, g^2, g^3, \ldots$ cannot all be different. Suppose, then, that $g^m = g^n$, with $m > n$. It then follows (by cancelling n gs from each side of this equation) that $g^{m-n} = e$, where $m - n$ is a positive integer. Thus some power of g equals the identity element of G. The smallest positive integer k such that $g^k = e$, is called the **order** of g, and this number is denoted by $o(g)$. If G is infinite it is also possible that for some positive integer k, $g^k = e$, in which case the smallest such k is also called the order of g. However, when G is infinite it usually happens that all the powers of g are different, in which case we say that g is an element of **infinite order**.

Since $e^1 = e$, the identity element of a group has order 1, and, clearly, in every group, this is the only element of order 1.

You should note that we are now using *order* in a new sense. Originally we used it to mean the number of elements in a group. It is important not to confuse these two meanings. They are, however, as you might expect, closely related. We will soon see that there is a close connection between this new meaning of *order* and the original meaning. First, however, it will be useful to prove the following Lemma.

Lemma 7.12
If g is an element of the group G with order k, then for each positive integer n,

$$g^n = e \Leftrightarrow n \text{ is a multiple of } k.$$

Proof
Since $o(g) = k$, $g^k = e$. Now suppose that n is a multiple of k, say $n = mk$, where m is a positive integer. Then

$$g^n = g^{mk} = (g^k)^m = e^m = e.$$

For the converse, suppose that $g^n = e$, but that n is not a multiple of k. Let r be the remainder when n is divided by k, so that $0 < r < k$, and for some positive integer m, $n = mk + r$. It follows that

$$e = g^n = g^{mk+r} = g^{mk}g^r = eg^r = g^r.$$

The fact that $g^r = e$, with $0 < r < k$, contradicts the definition of k as the least positive integer with $g^k = e$. This contradiction arose because we supposed that n was not a multiple of k. Thus n must be a multiple of k.

Note that in the course of this proof we have shown that if g has order k, then for each positive integer n, $g^n = g^r$, where r is the remainder when n is divided by r. Our proof covered the case where n is not a multiple of k, but if we extend our power notation so that $g^0 = e$, this is also true when n is a multiple of k. It follows that there are only k distinct powers of g, namely the elements of the set

$$\{e, g, g^2, g^3, \ldots, g^{k-1}\}.$$

We now show that this set is, in fact, a subgroup of G.

Lemma 7.13
Let g be an element of the group G of order k. Then the set

$$\{e, g, g^2, g^3, \ldots, g^{k-1}\}$$

is a subgroup of G of order k.

Proof
The closure property certainly holds, since if we combine two powers of g we obtain another power of g. The identity element is in the set, so it remains only to show that every element of the set has an inverse. If $1 \leqslant i < k$, g^{k-i} is in the set, and as

$$g^i g^{k-i} = g^{k-i} g^i = g^k = e,$$

it is the inverse of g^i.

The subgroup $\{e, g, \ldots, g^{k-1}\}$ is called the **subgroup generated by** g, and is denoted by $\langle g \rangle$. Notice the order of this subgroup is k, that is, it is the same as the order of the element g. This is the promised connection between the two senses of order. We can now use Lagrange's Theorem to deduce the following:

Theorem 7.14 *(Corollary to Lagrange's Theorem)*
If G is a finite group, and $g \in G$, then the order of g is a divisor of the order of G.

Proof
By Lagrange's Theorem $\#(\langle g \rangle)$ is a divisor of $\#(G)$. We have seen that $o(g) = \#(\langle g \rangle)$, and the Theorem follows.

As with Lagrange's Theorem itself, this Corollary only deals with the *possible* orders of the elements of a group. Not all these possibilities need be realized. For example, as follows from Exercise 7.5.2 below, the group S_4 of order 24, contains elements of orders 1, 2, 3 and 4 but not of orders 6, 8, 12 or 24 even though these numbers are also divisors of 24.

Now that we know about the orders of group elements, we are ready to tackle the problem about shuffles with which we began this Chapter. We can see now that it is a problem about the orders of permutations.

Exercises

7.5.1. Calculate the orders of the elements of the symmetry group of a square.

7.5.2. Calculate the orders of the elements of S_4.

7.5.3. If g is an element of a group of order 2, $gg = e$, and hence g is its own inverse, that is, $g^{-1} = g$. Groups in which every element (other than the identity) has order 2 are very special. Prove that every such group is commutative.

7.6 THE ORDERS OF PERMUTATIONS

You will remember that we began this chapter by asking how many riffle shuffles are needed to return a pack of cards to its original starting position. We can now rephrase this question in more technical language in terms of permutation groups.

A riffle shuffle corresponds to a certain permutation of a pack of 52 cards, and so we can regard it as an element of the permutation group S_{52}. Indeed, every way of shuffling a pack of cards corresponds to an element of this group. Let σ be some shuffle. The effect of performing this shuffle k times is given by the permutation σ^k. This has the effect of returning the pack to its original position provided that σ^k has no effect at all on the order of the cards, that is, if $\sigma^k = e$. Hence the number of times we need to repeat the shuffle σ to return the pack to its original position is the order of the permutation σ. Hence to solve our problem we need to know how to work out the orders of permutations and it is to this that we now turn.

It is when it comes to working out the order of a permutation that the cycle notation shows to real advantage. Let us begin with a simple example, by considering the permutation

$$\sigma = (1\ 2\ 3\ 4\ 5\ 6)$$

from S_6 which consists of a single cycle of length 6. We can represent this permutation by the following diagram:

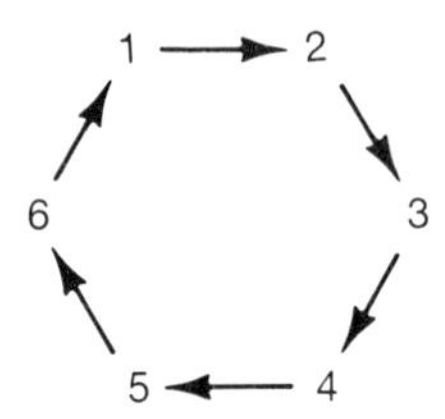

From the diagram it is clear that we have to repeat σ exactly 6 times to take the numbers back to their starting positions. That is, the order of σ is 6. Clearly, this would be the same for any cycle of length 6, and, more generally, we have the following result.

Lemma 7.15

If the permutation σ consists of a single cycle of length k, then the order of σ is k.

I introduced this lemma with the word 'clearly', which is my way of indicating that it should not require any proof.

Next we consider a permutation which involves more than one cycle. Let τ be the permutation

$$(1\ 2\ 3)(4\ 5)(6\ 7)$$

from S_7. How many times do we have to repeat τ to get back to the starting position? τ permutes the numbers 1, 2 and 3 in a cycle of length 3. So τ^k returns these numbers to their original positions provided that k is a multiple of 3. Similarly τ permutes 4 and 5 in a cycle of length 2, and therefore τ^k returns 4 and 5 to their starting positions provided that k is a multiple of 2. The same holds for the numbers 6 and 7 which also make up a cycle of length 2. It follows that

$$\tau^k = e \Leftrightarrow k \text{ is a multiple of 3 and a multiple of 2.}$$

Hence, the order of τ is the smallest positive integer which is both a multiple of 3 and a multiple of 2. So the order of τ is the least common multiple of 3 and 2, namely, 6. Again, it should be clear that the same is true for any permutation whose disjoint cycle decomposition consists of one cycle of length 3 and two cycles of length 2. More generally, the order of a permutation is the least common multiple of the lengths of the cycles in its disjoint cycle decomposition. Here is another example. The permutation

$$(1\ 2\ 3\ 4)(5\ 6)(7\ 8\ 9\ 10\ 11\ 12)$$

has order 12, because 12 is the least common multiple of 4, 2 and 6, which are lengths of the individual cycles.

We have thus seen that the order of a permutation is determined by the structure of its disjoint cycle decomposition. We call this structure the **cycle type** of the permutation, and we introduce an algebraic notation to represent this structure. This algebraic notation will play an important role in the subsequent chapters of this book.

The permutation $(1\,2\,3)(4\,5)(6\,7)$ has the same cycle type as the permutation $(4\,5\,7)(1\,6)(2\,3)$, as each consists of two cycles of length 2 and one cycle of length 3. We represent this cycle type by the algebraic symbolism

$$x_2^2 x_3^1.$$

In general the cycle type of a permutation σ is defined to be

$$x_{l_1}^{k_1} x_{l_2}^{k_2} \ldots x_{l_s}^{k_s}$$

if in the disjoint cycle decomposition of σ there are, for $1 \leqslant i \leqslant s$, k_i cycles of length l_i. Each k_i will be a positive integer, and the l_is will form an increasing sequence, that is, we will have $l_1 < l_2 < \ldots < l_s$.

Usually we omit the exponent k_i when it is equal to 1. So, for example, normally we will write the cycle type of $(1\,2\,3)(4\,5)(6\,7)$ as $x_2^2 x_3$. Do note carefully that although when writing a permutation in disjoint cycle form we do not bother to write down cycles of length 1, these 1-cycles must *not* be omitted when writing down the cycle type of a permutation. Thus, for example, the cycle type of the permutation

$$(1\ 2\ 3)(4\ 6)$$

from S_7 is

$$x_1^2 x_2 x_3,$$

since this permutation involves the two cycles (5) and (7) each of length 1 that we did not bother to write down above. Thus the identity element of S_n, whose disjoint cycle decomposition consists of n cycles of length 1, has cycle type x_1^n.

We can now express the general result for the order of a permutation as follows.

Lemma 7.15
If the permutation σ has cycle type $x_{l_1}^{k_1}x_{l_2}^{k_2} \dots x_{l_s}^{k_s}$ then the order of σ is the least common multiple of $l_1, l_2, \dots, l_s$ (which we write as $\mathrm{lcm}\{l_1, l_2, \dots, l_s\}$).

Suppose σ is the permutation from S_n whose cycle type is given by the expression $x_{l_1}^{k_1}x_{l_2}^{k_2} \dots x_{l_s}^{k_s}$. Then n different numbers occur in the cycles which make up the disjoint cycle decomposition of σ. The cycle type tells us that σ is made up of k_1 cycles of length l_1, k_2 cycles of length l_2 and so on. Hence we must have

$$k_1l_1 + k_2l_2 + \dots + k_sl_s = n.$$

This equation corresponds to a partition of n, with k_1 parts of size l_1, k_2 parts of size l_2 and so on. Hence the number of different cycles types of the permutations in S_n is $p(n)$, the number of partitions of n, which we discussed in some detail in Chapter 3.

Thus to obtain the orders of the elements of S_n we need only enumerate the partitions of n and, using Lemma 7.15, evaluate the least common multiples of the numbers which make up each partition. Thus for S_4 we have:

Partition	*Order*
4	4
3 + 1	3
2 + 2	2
2 + 1 + 1	2
1 + 1 + 1 + 1	1

Thus, as we remarked in the previous section, S_4 is a group of order 24 which has elements of orders 1, 2, 3 and 4 but not of orders 6, 8, 12 or 24.

To find out how many times we need to repeat an over riffle shuffle, or an under riffle shuffle to return a pack of cards to its original position, we need only work out the cycle type of the permutations corresponding to these shuffles. To find the best shuffle from this point of view amounts to finding the largest order of any element of S_{52}. We leave this as an exercise for the reader.

Exercises

7.6.1. Calculate the orders of the over riffle shuffle and the under riffle shuffle.

7.6.2. Find the largest order of the permutations in S_n, for $1 \leqslant n \leqslant 10$.

7.6.3. Find the largest order of the permutations in S_{52}. (Hint: $p(52)$ is so large that it is not practicable to tackle this question by listing all the partitions of 52. Your search for a partition which corresponds to a permutation of highest possible order might be guided by your answers to Exercise 7.6.2, which shows that only partitions made up of numbers of a special form need be considered.)

8

Group actions

8.1 COLOURINGS

We begin with a problem which leads us to the mathematics in this chapter.

Problem 1
How many different ways are there to colour the squares of a chessboard using the colours black and white?

The standard chessboard consists of a square 8×8 array of smaller squares, making 64 small squares in all. On a conventional chessboard these squares are coloured alternately black and white as shown, but clearly they could be coloured in many other ways. The problem is to work out exactly how many of these there are.

For the sake of simplicity we begin with the simpler problem of a 2×2 chessboard with only 4 squares. This will guide us to the general theory and so will enable us to solve Problem 1.

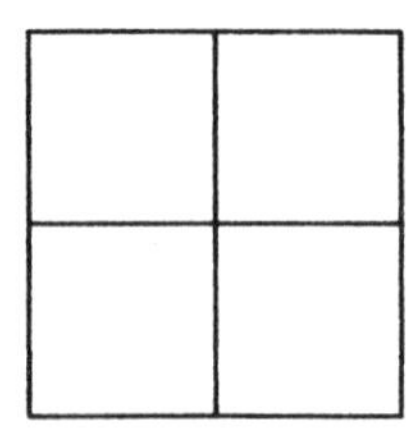

Since we have a choice of two colours for each square, and there are four squares to be coloured, it follows that the total number of colourings is $2^4 = 16$. These 16 colourings are shown in Figure 8.1. We have labelled them C1, C2, . . ., C16 for future reference.

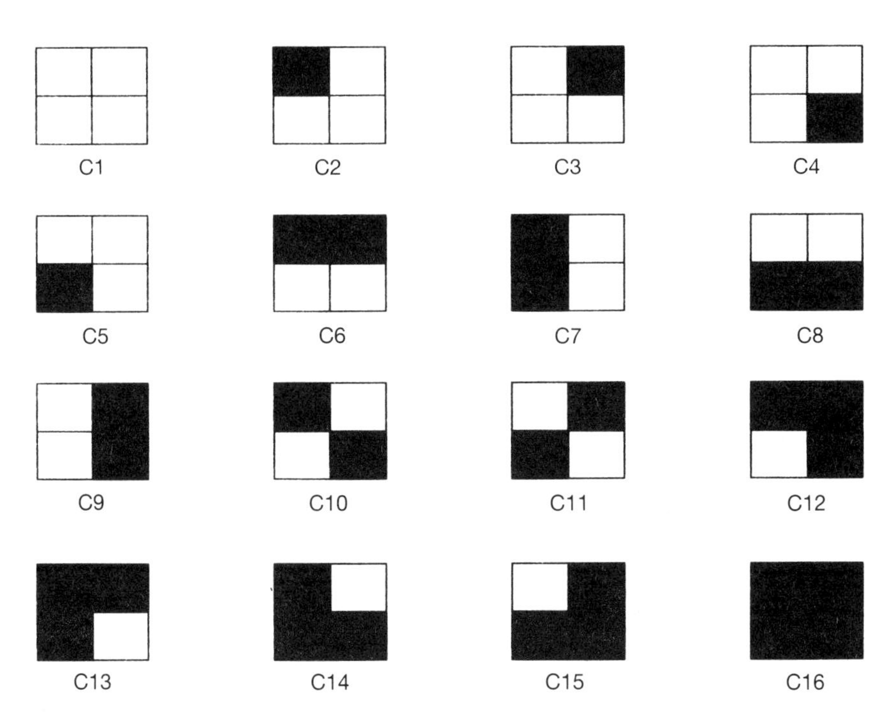

Fig. 8.1 The colourings of a 2×2 chessboard, using only black and white.

Are these colourings really all different? This depends on what we mean by 'different'. It does seem reasonable to say that some of these colourings really are the same, and that they only look different because the chessboard has been turned a different way. For example, C2 is really the same as C3, since we can get from C2 to C3 by giving the board a quarter turn clockwise.

From this point of view there are only six different patterns among the 16 different colourings. That is, we can put the colourings into the following sets, with all the colourings in the same set having the same pattern.

{C1}

{C2, C3, C4, C5}

{C6, C7, C8, C9}

{C10, C11}

{C12, C13, C14, C15}

{C16}

We can see that lurking in the background are the symmetries of a square. The reason why, for example, we have put C6 and C8 in the same set is that there is a symmetry of the square which takes us from one colouring to the other. Actually, in this case there are two symmetries which do this, the half-turn, and the reflection in the horizontal axis.

As it happens, the case of a 2×2 chessboard is so simple that it makes no difference whether we take into account reflections or not. Whenever there is a reflectional symmetry taking us from one colouring to another, there is also a rotational symmetry which does the same job. However, as soon as we move to 3×3 chessboards, this does make a difference. Consider, the following two colourings:

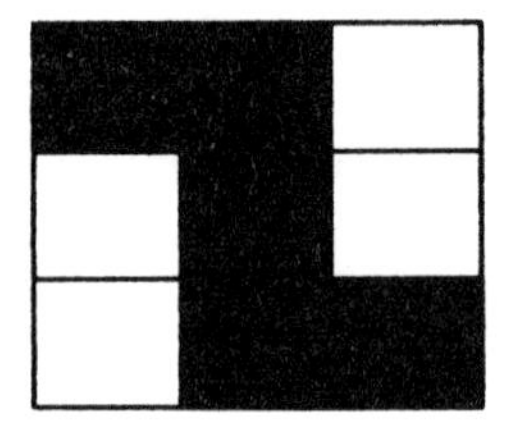

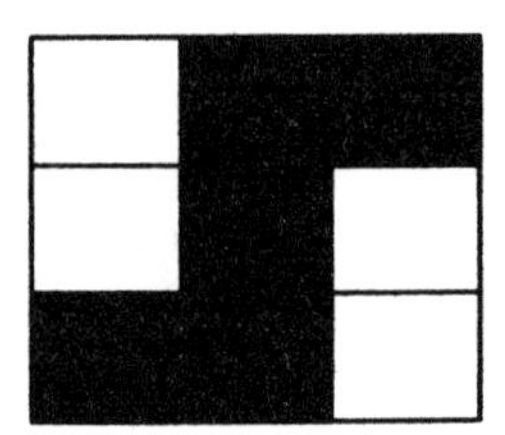

There is a reflectional symmetry which takes us from one to the other, but not a rotational symmetry. We are free to choose whether we wish to regard these colourings as having the same pattern or not. This amounts to deciding which group of symmetries we are going to use, either the full group of all eight symmetries of the square, or just the subgroup consisting of the rotational symmetries (with the identity, of course). The underlying theory is the same whichever group we choose.

The theory we are speaking about here deals with the interaction between a group (in our example the symmetries of a square) and the members of some other set (here the 16 different colourings of the chessboard). We begin to describe this theory in the next section.

Exercises

8.1.1. How many different colourings are there of a standard 8×8 chessboard, using two colours?

8.1.2. Give an example of two different colourings of a 2×2 chessboard, using *three* colours, which have the same pattern if we allow reflections, but not if we only consider rotational symmetries of the square.

8.2 THE AXIOMS FOR GROUP ACTIONS

We want to describe the general situation where we have an interaction between a group G and a set X. You may find it helpful to keep in mind throughout this general discussion the particular example concerning the symmetries of a square and colourings of a chessboard, described in section 8.1. Indeed, I will often refer to this example myself.

The symmetries of the square, in our example, interact with each colouring to produce a (possibly) different colouring. For example, the quarter-turn clockwise interacts with C2 to yield C3. The general situation will thus be that the interaction of an element of the group G with any member of the set X produces another member of X (possibly the same as the original one). In order to develop a general theory we shall require that this interaction satisfies a couple of simple properties. We are thus led to the following definition.

Definition

Let G be a group and let X be a set. We say that G **acts on** X if for each $g \in G$ and each $x \in X$, there is defined an element $g \cdot x \in X$ in such a way that the following properties hold.

GA1. For each $x \in X$, $e_G \cdot x = x$, where e_G is the identity of G.

GA2. For all $g_1, g_2 \in G$, and each $x \in X$,

$$g_1 \cdot (g_2 \cdot x) = (g_1 g_2) \cdot x.$$

You will note that GA1 says that the action of the identity element of G is always trivial; it fixes every element of X. Axiom GA2 relates the group action and the operation which defines the group. As Exercise 8.2.2 shows, it can be interpreted as expressing the fact that the mapping which associates each element of G with its action on X, is a **homomorphism** from G to $S(X)$. (A homomorphism from a group $(G, *)$ to a group $(H, \odot)$ is a mapping $\phi: G \to H$ such that for all $g_1, g_2 \in G$, we have $\phi(g_1 * g_2) = \phi(g_1) \odot \phi(g_2)$.)

Whenever a group acts on a set we say that we have a **group action**. We call GA1 and GA2 the **axioms for a group action**. We give some examples of group actions in just a moment. Before doing so, I need to stress that a group action is rather different from most of the algebraic operations that you will have previously met. In a group action the elements of the group G and of the set X can be completely different sorts of objects. Thus the combination of g and x to yield $g \cdot x$ is very different from, for example, the combination of two group elements g and h to produce $g * h$. Thus, although we often omit the symbol corresponding to a group operation, the symbol $\cdot$ which represents the group action *should never be left out*. It helps to remind us of the disparity between g and x.

Example 1
Let G be the group of symmetries of the square, and let X be the set of colourings of a 2×2 chessboard (see Figure 8.1). $g \cdot x$ is the colouring which results when the symmetry g is applied to the square colouring x (as in section 8.1). You should satisfy yourself that it does indeed satisfy the axioms for a group action.

Example 2
Let G be the group of rotational symmetries of the cube, and let X be the set of colourings of the cube using the three colours red, white and blue. The action $g \cdot x$ is defined as in Example 1. We will be discussing this example in some detail later in this chapter. Clearly there are many other examples of a similar kind.

Example 3
Let $G = (\mathbb{R}, +)$ be the group of real numbers with the operation of addition. Let $X = \mathbb{R}^2$. (Since the elements of X are vectors, we will write them in bold type.) The action is defined by

$g \cdot \mathbf{x}$ = the point to which $\mathbf{x}$ is moved by a rotation through the angle g about the origin.

It is not difficult to check that the axioms for a group action hold in this case. The identity element of G is the number 0. Clearly a rotation through an angle 0 leaves every point fixed. So for all $\mathbf{x} \in \mathbb{R}^2$, $0 \cdot \mathbf{x} = \mathbf{x}$. Hence GA1 is satisfied. To see that GA2 is satisfied note that in the context of this example it says that for all $g, h \in \mathbb{R}$, and each $\mathbf{x} \in \mathbb{R}^2$, $g \cdot (h \cdot \mathbf{x}) = (g + h) \cdot \mathbf{x}$. In other words, rotation through an angle h, followed by rotation through an angle g, has the same effect on $\mathbf{x}$ as rotation through the angle $g + h$, and clearly this is true.

Example 4
Let G be any group, and let X be the set of elements of G. We define the group action by

$$\text{for } g, x \in G \qquad g \cdot x = gxg^{-1}.$$

This is a more sophisticated example than the three above, since it is a case of a group acting on itself. The action in this case is usually called

conjugation. The group element gxg^{-1} is called the **conjugate of x by g**.

We check that conjugation does indeed define a group action. Since, for each $x \in G$,

$$e \cdot x = exe^{-1} = exe = x,$$

GA1 is satisfied. Now suppose $g_1, g_2 \in G$ and $x \in X$, then

$$\begin{aligned} g_1 \cdot (g_2 \cdot x) = g_1(g_2 \cdot x)g_1^{-1} &= g_1(g_2 x g_2^{-1})g_1^{-1} \\ &= (g_1 g_2)x(g_2^{-1} g_1^{-1}) \\ &= (g_1 g_2)x(g_1 g_2)^{-1} \\ &= (g_1 g_2) \cdot x, \end{aligned}$$

which shows that GA2 also holds.

Conjugation plays an important role in group theory, but it is not greatly important from the point of view of this book. None the less, for the benefit of readers who are already familiar with it we shall, from time to time, draw attention to those properties of this operation that we can easily deduce from the theory of group actions.

We end this section with a simple lemma about group actions which we will need later on.

Lemma 8.1

Let G be a group which acts on the set X. Then for each $g \in G$, and all $x, y \in X$,

$$g \cdot x = y \Leftrightarrow g^{-1} \cdot y = x.$$

Proof

Suppose that $g \cdot x = y$. Then

$$\begin{aligned} g^{-1} \cdot y &= g^{-1} \cdot (g \cdot x) \\ &= (g^{-1} \cdot g) \cdot x \qquad \text{by GA2} \\ &= e \cdot x \\ &= x \qquad \text{by GA1.} \end{aligned}$$

The converse implication is proved in a similar way.

Exercises

8.2.1. In each of the following cases determine whether the group action axioms hold.

(a) $G = (\mathbb{R}^*, \times)$, $X = \mathbb{R}^2$, $g \cdot (x, y) = (gx, gy)$.

(b) $G = (\mathbb{R}, +)$, $X = \mathbb{R}^2$, $g \cdot (x, y) = (gx, gy)$.

(c) $G = \left\{ \begin{pmatrix} a & 0 \\ c & b \end{pmatrix} : a, b, c \in \mathbb{R}, ab \neq 0 \right\}$,

$$X = \mathbb{R}^2, \begin{pmatrix} a & 0 \\ x & b \end{pmatrix} \cdot (x, y) = (ax, cx + by).$$

8.2.2. Let G be a group which acts on the set X. For each $g \in G$, let $\phi_g : X \to X$ be the mapping defined by

$$\phi_g(x) = g \cdot x.$$

(a) Show that for each $g \in G$, the mapping ϕ_g is a permutation of X.

(b) It follows from (a) that for each $g \in G$, $\phi_g \in S(X)$, where $S(X)$ is the set of all permutations of X. We know that $S(X)$ forms a group with respect to the operation of composition of mappings. Show that the mapping $\Phi : G \to S(X)$ given by

$$\Phi(g) = \phi_g$$

is a group homomorphism, i.e. that for all $g, h \in G$, $\Phi(gh) = \Phi(g) \circ \Phi(h)$.

8.3 ORBITS

We came across group actions when we were trying to count the number of different patterns that can be obtained by colouring the squares of a chessboard. The relevant group action was that of the group of symmetries of a square on the different colourings. We decided that two colourings were essentially the same if the action of one of these symmetries on one colouring produced the other. We can now spell out this definition in general terms using the language of group actions.

Definition
Suppose G is a group which acts on the set X. We define a relation $\sim_G$ on X as follows:

$$\text{For all } x, y \in X, x \sim_G y \Leftrightarrow \text{for some } g \in G, g \cdot x = y.$$

As the notation suggests, $\sim_G$ is an equivalence relation on X, and this we now prove.

Lemma 8.2
$\sim_G$ is an equivalence relation on the set X.

Proof
We need to prove that the relation $\sim_G$ is reflexive, symmetric and transitive. Notice from the proof that these properties follow from the identity, inverses and closure properties of a group, respectively, making use of the axioms for a group action.

Suppose $x \in X$. Now, G has an identity element e, and by GA1,

$e \cdot x = x$. It follows that $x \sim_G x$, and hence that the relation $\sim_G$ is reflexive.

Suppose, next, that $x, y \in X$ and $x \sim_G y$. So there is some $g \in G$ with $g \cdot x = y$. As G is a group, it contains also g^{-1}, and by Lemma 8.1, $g^{-1} \cdot y = x$. Hence $y \sim_G x$. This shows that $\sim_G$ is a symmetric relation.

Finally, suppose that $x, y, z \in X$, $x \sim_G y$ and $y \sim_G z$. Then there exist $g, h \in G$ such that $g \cdot x = y$ and $h \cdot y = z$. Since G is a group, $hg \in G$, and, using GA2,

$$hg \cdot x = h \cdot (g \cdot x) = h \cdot y = z.$$

Hence $x \sim_G z$. This shows that $\sim_G$ is transitive. This completes the proof that $\sim_G$ is an equivalence relation.

In our example of the colourings of a chessboard, the equivalence classes, with respect to this equivalence relation, are those sets of colourings which all have the same pattern. Thus our problem about how many different patterns there are, can be restated as: 'How many different equivalence classes are there?'

In Example 2 in section 8.2, the example involving colourings of a cube, the equivalence classes also consist of colourings of the same pattern. It can be seen that in Example 3 the equivalence classes consist of circles centred on the origin, since there is a rotation about the origin taking $\mathbf{x}$ to $\mathbf{y}$ if and only if $\mathbf{x}$ and $\mathbf{y}$ are at the same distance from the origin.

In Example 4, where the group action is conjugation, the equivalence classes are the **conjugacy classes** of the group in question.

In general, the equivalence classes of the relation $\sim_G$ are called the **orbits** of the group action. The orbit to which the element $x \in X$ belongs is written as O_x. Thus for $x, y \in X$,

$$O_x = O_y \Leftrightarrow x \sim_G y.$$

Since $x \sim_G y$ if and only if $y = g \cdot x$ for some $g \in G$, it follows that

$$O_x = \{g \cdot x : g \in G\}.$$

Lemma 8.3

As the orbits are equivalence classes they partition X into disjoint sets. Our question about the number of patterns that we get when we colour chessboards, is thus the question of how many different orbits there are. Before we can give the theorem which answers this question we need to develop one more theoretical idea, which we do in the next section.

Exercise

8.3.1. For each of the examples of a group action, given in the Exercise 8.2.1, find the orbits of the action.

8.4 STABILIZERS

Whenever we have a group action, with, say, a group G acting on a set X, the identity element of G fixes every element of X, in the sense that $e{\cdot}x = x$. This is built into the definition of a group action via axiom GA1. Other elements of G may fix certain members of X. For example, if we go back to the colourings of a chessboard, (Figure 8.1), we see that the diagonal reflection r fixes C1, C2, C4, C10, C11, C13, C15 and C16, since all these colourings are symmetrical about the relevant diagonal, and so are unchanged when we reflect in it.

The type of symmetry of a particular colouring is determined by those elements of G which fix it. We call the set of those group elements of G which fix a particular $x \in X$, the **stabilizer** of x, and we denote it by S_x. Thus

$$S_x = \{g \in G : g{\cdot}x = x\}.$$

Again, looking at our example of colourings of chessboards, you will see that the stabilizer of the colouring C1 is $\{e, r\}$ and that of C10 is $\{e, b, r, s\}$. These stabilizers are both subgroups of G. The next lemma tells us that this is not an accident.

Lemma 8.4

If the group G acts on a set X, then for each $x \in X$, S_x is a subgroup of G.

Proof

We check the subgroup conditions. First, suppose that $g, h \in S_x$. Then $g{\cdot}x = x$ and $h{\cdot}x = x$. Therefore, using GA2,

$$gh{\cdot}x = g{\cdot}(h{\cdot}x) = g{\cdot}x = x,$$

and hence $gh \in S_x$. Hence the closure condition is satisfied.

By GA1, $e{\cdot}x = x$, and hence S_x contains the identity element of G. Finally suppose $g \in S_x$. Then $g{\cdot}x = x$, and so, by Lemma 8.1, $g^{-1}{\cdot}x = x$, and hence $g^{-1} \in S_x$. So S_x contains the inverse of each of its elements. This completes the proof that it is a subgroup of G.

We are interested in using stabilizers to help solve combinatorial problems. They are of interest to group theorists because, as this Lemma shows, they help identify subgroups of a given group.

Once again, it is instructive to return to our example of the colourings of the 2×2 chessboard, and to list the orbits and the stabilizers of the elements of X, which we do in Table 8.1.

The pattern is clear. The larger the orbit, the smaller is the stabilizer and vice versa. Indeed there is a very close relationship between the size of the orbit and the size of the stabilizer. In each

Table 8.1 Orbits and stabilizers of the colourings of a 2 × 2 chessboard

Element of X	*Orbit*	*Stabilizer*
C1	{C1}	$\{e, a, b, c, p, q, r, s\}$
C2	{C2, C3, C4, C5}	$\{e, r\}$
C3		$\{e, s\}$
C4		$\{e, r\}$
C5		$\{e, s\}$
C6	{C6, C7, C8, C9}	$\{e, q\}$
C7		$\{e, p\}$
C8		$\{e, q\}$
C9		$\{e, p\}$
C10	{C10, C11}	$\{e, b, r, s\}$
C11		$\{e, b, r, s\}$
C12	{C12, C13, C14, C15}	$\{e, s\}$
C13		$\{e, r\}$
C14		$\{e, s\}$
C15		$\{e, r\}$
C16	{C16}	$\{e, a, b, c, p, q, r, s\}$

case $\#(O_x) \times \#(S_x) = 8$, which is the number of elements in the group G. Again, this is not a coincidence, but an important result which holds in all cases.

Theorem 8.5 *(The Orbit–Stabilizer Theorem)*
Let G be a group which acts on a set X. Then for each $x \in X$,

$$\#(O_x) \times \#(S_x) = \#(G). \tag{8.1}$$

Proof
By Lemma 8.4, S_x is a subgroup of G, and by Lagrange's Theorem (Theorem 7.11),

$$|G : S_x| \times \#(S_x) = \#(G). \tag{8.2}$$

Comparing equations (8.1) and (8.2), we see that we need to show that

$$\#(O_x) = |G : S_x|. \tag{8.3}$$

The elements of O_x have the form $g{\cdot}x$, for $g \in G$, and the cosets of S_x have the form gS_x for $g \in G$. So we can prove equation (8.3) by showing that the correspondence $g{\cdot}x \leftrightarrow gS_x$ is a one-to-one correspondence between the elements of O_x and the cosets of S_x in G. In other words, we need to show that for all $g_1, g_2 \in G$,

$$g_1 S_x = g_2 S_x \Leftrightarrow g_1 \cdot x = g_2 \cdot x.$$

This we now do. Suppose $g_1, g_2 \in G$. Then

$$\begin{aligned} g_1 S_x = g_2 S_x &\Leftrightarrow g_2^{-1} g_1 \in S_x, && \text{by Lemma 7.7,} \\ &\Leftrightarrow (g_2^{-1} g_1) \cdot x = x, && \text{by the definition of } S_x, \\ &\Leftrightarrow g_2^{-1} \cdot (g_1 \cdot x) = x, && \text{by GA2,} \\ &\Leftrightarrow g_1 \cdot x = g_2 \cdot x, && \text{by Lemma 8.1.} \end{aligned}$$

From Theorem 8.5 we can easily derive the following:

Corollary 8.6

Let G be a finite group which acts on the set X. Then the number of elements in each orbit is a divisor of the order of G.

Proof

This corollary is an immediate consequence of equations (8.1).

If we apply Corollary 8.6 to the case where the group action is conjugation, we can deduce that the number of elements in a conjugacy class is a divisor of the number of elements in the group.

The Orbit–Stabilizer Theorem relates the number of elements in each orbit to the number of elements in each stabilizer, but as the next theorem shows, it can very easily be rearranged to give a formula for the number of different orbits.

Theorem 8.7

Let G be a finite group which acts on a set X. Then the number of distinct orbits is

$$\frac{1}{\#(G)} \sum_{x \in X} \#(S_x).$$

Proof

Suppose that there are k distinct orbits $O_{x_1}, \ldots, O_{x_k}$. For $1 \leqslant t \leqslant k$,

$$\begin{aligned} \sum_{x \in O_{x_t}} \#(S_x) &= \sum_{x \in O_{x_t}} \frac{\#(G)}{\#(O_x)} && \text{by the Orbit–Stabilizer Theorem,} \\ &= \sum_{x \in O_{x_t}} \frac{\#(G)}{\#(O_{x_t})} && \text{as } O_x = O_{x_t} \text{ for each } x \in O_{x_t} \\ &= \frac{\#(G)}{\#(O_{x_t})} \times \#(O_{x_t}) \\ &= \#(G). \end{aligned} \tag{8.4}$$

Since the sum of the numbers $\#(S_x)$ taken over the elements of just one orbit comes to $\#(G)$ in each case, when we take the sum over all the elements of X we get a total of $\#(G)$ for each orbit and hence $k\#(G)$ in all. That is, in symbols,

$$\begin{aligned}\sum_{x \in X} \#(S_x) &= \sum_{1 \leqslant t \leqslant k} \sum_{x \in O_{x_t}} \#(S_x) \\ &= \sum_{1 \leqslant t \leqslant k} \#(G) \qquad \text{by equation (8.4),} \\ &= k\#(G).\end{aligned}$$

Therefore

$$k = \frac{1}{\#(G)} \sum_{x \in X} \#(S_x).$$

This theorem gives us a formula for the number of orbits, but is it a useful formula? It is easy to use in our standard example of the colourings of the 2×2 chessboard. We get the following table of values of $\#(S_x)$.

x	C1	C2	C3	C4	C5	C6	C7	C8	C9	C10	C11	C12	C13	C14	C15	C16
$\#(S_x)$	8	2	2	2	2	2	2	2	2	4	4	2	2	2	2	8

$\Sigma_{x \in X} \#(S_x) = 48$, and so the number of distinct orbits is

$$\frac{1}{\#(G)} \sum_{x \in X} \#(S_x) = \frac{1}{8} \times 48 = 6,$$

which agrees with the value we have already found.

This example is, however, deceptively simple. We were only able to use the formula given by Theorem 8.7 because we could easily list the elements of X. However, because we could easily list them, we did not need any elaborate theory to work out the number of different patterns. We were able to count them directly. In the case we are really interested in, that of 8×8 chessboards, this method would be completely impractical. The number of different colourings of an 8×8 chessboard, using just two colours, is 2^{64}, which is greater than 10^{19}. So it is completely impractical to sum directly the terms $\#(S_x)$ for each $x \in X$.

We can get round this difficulty if we note that, although the number of colourings is vastly greater in this case, the underlying group G remains the same. So if we can replace the sum over X by sum over G, the 8×8 chessboard is no harder to deal with than the 2×2 chess board. Before showing how to do this, we make a digression in the next chapter, by giving an example where Theorem 8.7 can be applied directly. The application is in the theory of graphs, which is, in any case, of independent interest.

Exercises

(In these exercises we show how the theory of group actions can be used to derive some theorems about groups. We only scratch the surface, and you should consult a book about groups to learn more about this. See the Reading List at the end of this book for some suggestions for further reading in this area.)

8.4.1. The identity element of a group, G, commutes with every other element of G. There may or may not be other elements of G with this property. The **centre** of G, written $Z(G)$, consists of those elements of G which commute with every element of G. That is

$$Z(G) = \{g \in G: \text{for all } x \in G,\ gx = xg\}.$$

(a) Prove that $Z(G)$ is a subgroup of G.
(b) Prove that $g \in Z(G)$ if and only if the conjugacy class of g is $\{g\}$, i.e. if and only if g is conjugate to itself alone.
(c) Prove that if $\#(G)$ is a prime power, then G has a non-trivial centre. That is, prove that if for some prime number p and some positive integer k, $\#(G) = p^k$, then $\#(Z(G)) > 1$.

8.4.2. We have seen that the converse of Lagrange's Theorem is not, in general, true. However there are some special cases where it is possible to prove a converse of this theorem. In this Exercise we indicate how group actions can be used to prove the following theorem due to Cauchy.

Theorem 8.8

If G is a group and p is a prime number which is a divisor of $\#(G)$, then G contains an element of order p.

We let X be the set of all ordered p-tuples of elements of G whose product gives the identity element of G. That is,

$$X = \{(g_0, g_1, \ldots, g_{p-1}) \in G^p: g_0 g_1 \ldots g_{p-1} = e\}.$$

(a) Suppose that $\#(G) = n$. How many p-tuples are there in X?
We define an action of the group $\mathbb{Z}_p$ on X as follows:
For $k \in \mathbb{Z}_p$ and $(g_0, g_1, \ldots, g_{p-1}) \in X$,

$$k \cdot (g_0, g_1, \ldots, g_{p-1}) = (g_{0+k}, g_{1+k}, \ldots, g_{p-1+k}), \tag{8.5}$$

where the additions are carried out in $\mathbb{Z}_p$, i.e. modulo p.
(b) Prove that for all $k \in \mathbb{Z}_p$ and $(g_0, g_1, \ldots, g_{p-1}) \in X$,

$$k \cdot (g_0, g_1, \ldots, g_{p-1}) \in X.$$

(c) Prove that equation (8.5) does define a group action of $\mathbb{Z}_p$ on X.
(d) Prove that an element $(g_0, g_1, \ldots, g_{p-1})$ of X is fixed by every element of $\mathbb{Z}_p$ if and only if $g_0 = g_1 = \ldots = g_{p-1}$.

It follows from (d) that elements of X of the form $(g_0, g_0, \ldots, g_0)$ are in an orbit by themselves, and every other element of X belongs to an orbit containing more than one element. Now the p-tuple $(g_0, g_0, \ldots, g_0)$ is in X if and only if $g_0^p = e$. The p-tuple $(e, e, \ldots, e)$ is in X. For $g_0 \neq e$, the p-tuple $(g_0, g_0, \ldots, g_0)$ is in X if and only if $o(g_0) = p$. Hence to prove that G contains at least one element of order p we have only one more step to carry out.

(e) Prove that if n is divisible by the prime number p, there is more than one orbit containing just one element of X.

The remarks above show that when you have proved this last statement, you will have completed the proof of Cauchy's Theorem.

9

Graphs

9.1 WHAT ARE GRAPHS?

The word **graph** is used in mathematics in two quite distinct senses. You probably met it for the first time in the context of *graphs of functions*; the usual curves that we draw in the plane to represent the behaviour of real-valued functions. In the context of combinatorics, *graph* has a quite different meaning, which is now explained.

In combinatorics a graph consists of a set of points, called **vertices** (or, by some people, **nodes**), some of which are joined by lines, called **edges**. Here are some examples of graphs:

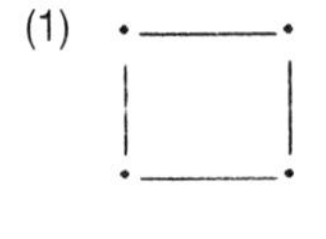

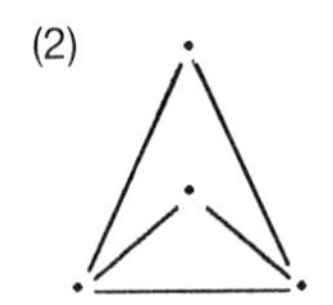

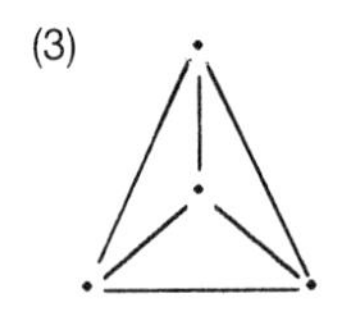

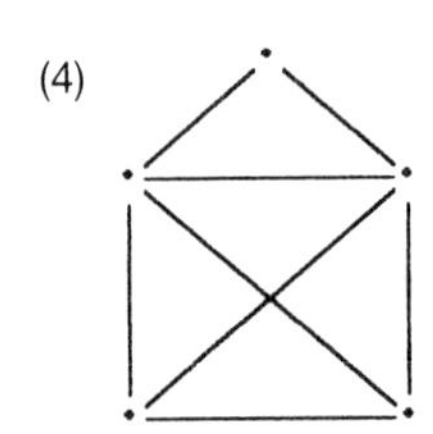

The convention is that when drawing pictures of graphs, we represent the vertices by dots. Points where the lines in our pictures cross are not necessarily vertices of the graph: in the fourth example above, there are only five vertices, and two edges cross at a point where there is no vertex of the graph.

These pictures are only informal representations of graphs. We can give a more formal mathematical definition which does not depend on pictorial representations. We specify a graph when we say which its vertices are, and which pairs of vertices are connected (by edges). Looked at this way, there is no reason why the vertices need to be geometrical points, nor the edges geometrical lines. So we will allow the vertices to be any set we like. The edges relate pairs of vertices. If the vertex u is connected to the vertex v, then v is also connected to u, and hence this relation of being connected is a symmetric relation. Since we choose not to allow the case where a vertex is connected to itself, this relation is also irreflexive. Thus we end up with the following formal definition.

Definition

A graph is a pair $\mathcal{G} = (V, E)$ where V is a set and E is an irreflexive symmetric relation on V. Two vertices $u, v \in V$ are **connected** in the graph if and only if u and v are related by the relation E. We write this as uEv. The graph is **finite** if the set, V, of vertices is finite. We will only be concerned with finite graphs, and in future we will not bother to stipulate specifically that the graphs with which we are dealing are all finite.

The disadvantage of this definition of a graph is that we seem to have lost sight of the edges. All we have is a set of vertices and the relation E. However, if we are given a graph (V, E), defined as above, we can always define an edge of the graph to be a two-element subset $\{u, v\}$ of V such that uEv. In some approaches a graph is defined so that E, instead of being a relation, consists of the set of two element subsets of V corresponding to the edges.

The definition of *graph* that we have just given is not the most general one we might have given. We have already noted that we are not allowing vertices to be connected to themselves. So we have ruled out what are usually known as **loops** in our graphs. Also, by interpreting the edges as being determined by a relation E, we are saying that any two vertices are either connected or not connected, but we are not allowing two vertices to be connected by more than one edge. In the jargon of graph theory, we are not allowing **multiple edges**.

Usually these two restrictions are not imposed from the outset. Graphs with no loops and no multiple edges are then called **simple graphs**. The restriction to simple graphs is a real restriction, but much of the theory of graphs does deal with this special case.

Graphs occur in many practical situations. For example a graph can represent a road or rail network. Graphs occur in scheduling and timetabling problems, and in critical path analysis. In some of these practical situations, another generalization of our definition of a graph becomes relevant. Often traffic can only move along a road in one particular direction. So in the theory of **directed graphs** each edge has associated with it a specific direction. Graphs are also the subject of a great deal of study for their own sake. For example, the well-known Four Colour Theorem is a theorem of graph theory. In section 9.2 we describe another application of graph theory, the application from which the subject acquired its name.

Because of its many applications and interesting problems graph theory is a large subject in its own right. We are not, however, going to discuss graph theory in general. Instead, we confine our attention to some counting problems involving graphs. For more information about graphs in general, you should consult one of the books in the Reading List at the end of this book.

The basic counting problem about graphs concerns how many different graphs there are with a given number of vertices. Before we can answer this question, we need to make clear what we mean by 'different graphs'. The following example will make this clear. For convenience, we continue to represent graphs by diagrams. We have labelled the vertices so that we can easily refer to them.

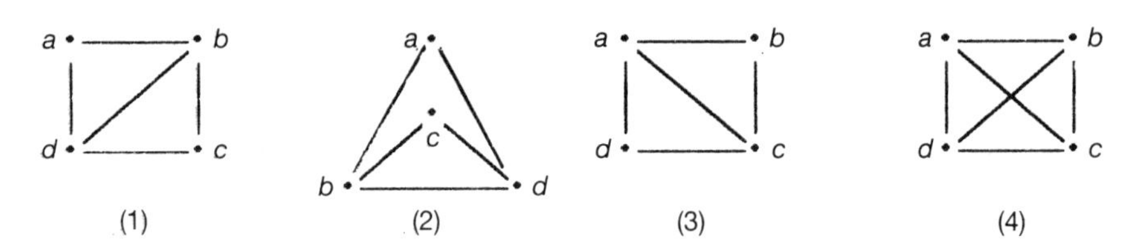

Diagrams (1) and (2) above look different, but on closer examination you will see that they are pictures of the same graph. The vertex set is $\{a, b, c, d\}$ in each case, and both pictures represent the same relation on this set. That is, in both pictures, a is connnected to b and d, b is also connected to c and d, and c to d, but a is not connected c.

The vertex set is again $\{a, b, c, d\}$ in diagram (3), but the relation is not the same. In (3) a is connected to c, whereas these vertices are not connected in (1). However, although the pictures are not exactly the same, one can be obtained from the other by a rotation. In other words if we rename the vertices in (3), the picture represents exactly the same relation as (1). All we have to do is to interchange a and b, and also c and d. The technical name for this renaming of the vertices is an **isomorphism**. The mapping φ which interchanges a and b, c and d, has the property that vertices x and y are connected in (1) if and only if $\varphi(x)$ and $\varphi(y)$ are connected in (3).

Definition

Let $\mathcal{G} = (V, E)$ and $\mathcal{G}' = (V', E')$ be two graphs. An isomorphism, φ, is a bijection $\varphi: V \to V'$ such that for all $u, v \in V$,

$$uEv \Leftrightarrow \varphi(u)E'\varphi(v).$$

The graphs $\mathcal{G}$ and $\mathcal{G}'$ are said to be **isomorphic** if there is an isomorphism between them. If $\mathcal{G}$ is isomorphic to $\mathcal{G}'$ we write

$$\mathcal{G} \simeq \mathcal{G}'.$$

Thus the mapping $\varphi : \{a, b, c, d\} \to \{a, b, c, d\}$ defined by

$$\varphi(a) = b,$$
$$\varphi(b) = a,$$
$$\varphi(c) = d,$$
$$\varphi(d) = c,$$

is an isomorphism between graphs (1) and (3) above.

The graph represented by diagram (4) is not isomorphic to the graph in diagram (3). This should be immediately obvious. In more complicated cases it can be difficult to tell whether or not two graphs are isomorphic. One simple method often works.

If v is a vertex of a graph $\mathcal{G}$, the **degree** of v, written $\delta(v)$, is the number of vertices which are joined to v. That is,

$$\delta(v) = \#(\{u \in V : vEu\}).$$

If $\varphi : V \to V'$ is an isomorphism, then for each $v \in V$,

$$\delta(v) = \delta(\varphi(v)).$$

Now in graph (4) every vertex has degree 3, while in graph (3) two vertices have degree 3 and two have degree 2. Hence there cannot be an isomorphism between them. Counting the degrees of the vertices can often be used to settle negatively the question of whether two graphs are isomorphic.

Because every edge has two vertices as its endpoints, the following holds.

Lemma 9.1

Let $\mathcal{G} = (V, E)$ be a graph. Then

$$\sum_{v \in V} \delta(v) = 2e, \qquad \text{where } e \text{ is the number of edges,}$$

and hence the sum is always an even number.

Since isomorphic graphs are essentially the same, our question about the number of different graphs can be rephrased as

How many non-isomorphic graphs are there with n vertices?

This turns out to be a rather difficult question, and we need to develop some more theory before we can begin to answer it. There is an easier version of this problem, dealing with **labelled graphs**, to which we turn in the next section.

Exercise

9.1.1. Although in general it is quite difficult to determine how many different graphs there are with n vertices, when n is small this can be settled by listing all the different graphs. For $1 \leqslant n \leqslant 4$, determine how many different graphs there are with n vertices.

9.2 LABELLED GRAPHS

To explain what labelled graphs are, we need to go back to the early days of graph theory, and to the problems from which the subject got its name.

Early in the nineteenth century it was realized that chemical substances are made up of elements. The properties of different substances could largely be explained by the different proportions of the elements which made them up. But in a few cases it was known that substances that seemed to be made up of the same elements in the same proportions nevertheless had different properties. Such substances are known as **isomers**. So knowing the constituents of a substance was not enough. It appeared to be relevant to work out how the different elements were combined to form molecules.

Various different notations were devised to represent the arrangement of atoms in molecules. The system which led to the method in common use today seems to have originated with Alexander Crum Brown, a chemist at the University of Edinburgh. He described his system as providing a 'graphic notation' for molecules. His 'graphic notation' has become our 'graphs'.

I have said that this notation 'seems' to have originated with Alexander Crum Brown, because questions of priority are notoriously difficult to settle, and views on who first had a particular idea vary a good deal and especially from one country to the next. My information comes from an anthology of historic papers on graph theory.*

Anyway, here is an example of Crum Brown's graphic notation, used to show two isomers of compounds with the same chemical formula C_3H_7OH. In the original notation, Crum Brown put circles round the letters denoting atoms of particular elements, but in line with the modern notation, I have omitted these.

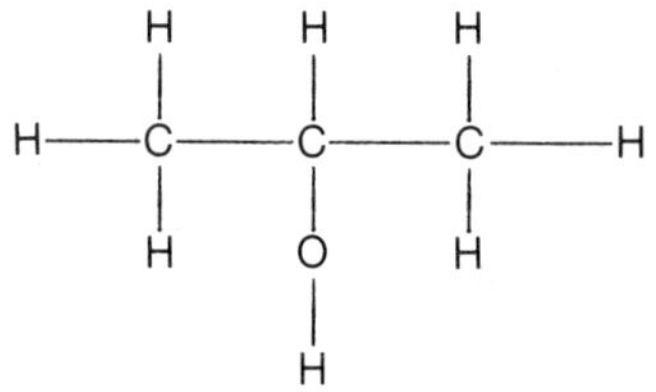

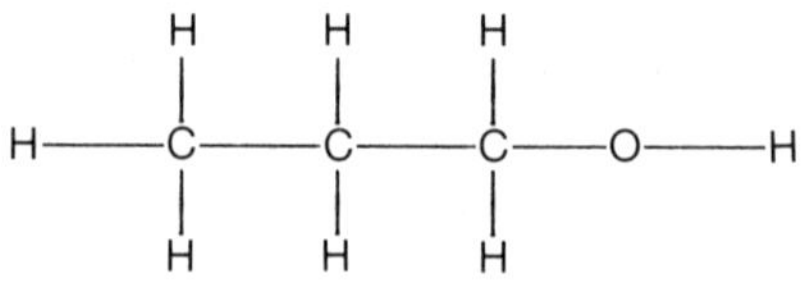

*Norman L. Biggs, E. Keith Lloyd and Robin J. Wilson, *Graph Theory 1736–1936*, Clarendon Press, Oxford, 1976.

These diagrams are very much like the pictures of graphs with which we began this chapter. They consist of vertices connected by edges. The only real difference is that in the present case each vertex is associated with a particular letter which represents an element. We can think of these letters as labelling the vertices, and thus the resulting structures are *labelled graphs*. In the case of labelled graphs corresponding to chemical elements we have the restriction that particular letters have to be attached to vertices of particular degrees, corresponding to the valency of the element. Also, in some molecules there are double valency bonds between atoms, for example in the well-known benzene ring and in the carbon dioxide molecule represented below,

O
‖
C
‖
O

so we have to allow multiple edges in these graphs. All this makes the problem of enumerating the number of possible isomers associated with a particular chemical formula rather difficult. So instead we shall concern ourselves with just one easy counting problem concerning labelled graphs.

Suppose that we have a graph $\mathcal{G}$ with k vertices. By a **labelling** of $\mathcal{G}$ we mean an assignment of the numbers $1, 2, \ldots, k$ to the vertices of $\mathcal{G}$. Thus, strictly, a labelling of $\mathcal{G}$ is a bijection $l : \{1, 2, \ldots, k\} \to V$. The problem we shall deal with is how many different labellings a given graph $\mathcal{G}$ has?

The following example will explain what we mean by 'different labellings' in this context. Here are three labellings of the same graph.

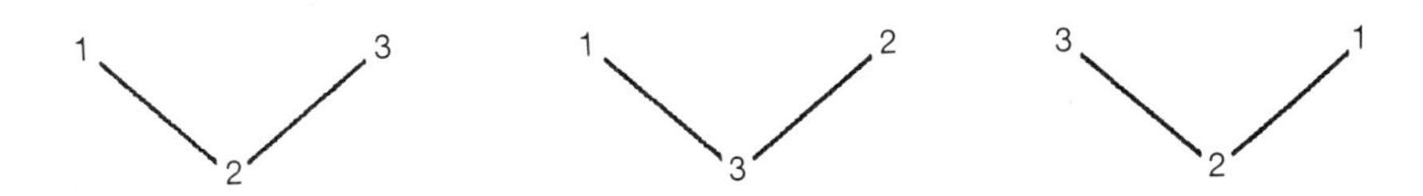

However, there is no real difference between the first and third labellings. In each case the vertex which is labelled 2 is connected to the other vertices, but the vertices labelled 1 and 3 are not connected. In other words the correspondence between the vertices which associates vertices with the same labels is an isomorphism from $\mathcal{G}$ to itself. Such an isomorphism from a graph to itself is called an **automorphism** of the graph.

On the other hand, the first and second labellings are clearly different. In the first the vertex of degree 2 is labelled with the number 2, while in the second it gets the label 3. Hence the correspondence between vertices with the same labels is certainly not an automorphism of the graph.

We can formulate our problem so that it fits the theory of group actions. Suppose we are given some fixed graph $\mathcal{G}$ with k vertices. The relevant group is the group of all the automorphisms of $\mathcal{G}$, which we write $\mathcal{A}(\mathcal{G})$. Yes, it is a group. This is straightforward to prove (Exercise 9.2.2), but for future reference we state it as a lemma.

Lemma 9.2

The set of all automorphisms of a particular graph $\mathcal{G}$ forms a group with composition (of mappings) as the operation.

Now we are ready to explain what the group action is. The relevant set is the set of all labellings of $\mathcal{G}$, which we will denote by $L(\mathcal{G})$. Since a labelling of $\mathcal{G}$ is a bijection from the set $\{1, 2, \ldots, k\}$ to V, the vertex set of $\mathcal{G}$, there are $k!$ different labellings in $L(\mathcal{G})$. The group $\mathcal{A}(\mathcal{G})$ of automorphisms of $\mathcal{G}$ acts on $L(\mathcal{G})$ in the obvious way. Indeed since a labelling, l, is a mapping

$$l : \{1, 2, \ldots, k\} \to V,$$

and an automorphism, θ, of G is a mapping

$$\theta : V \to V,$$

the group action is just composition of mappings, that is

$$\theta \cdot l = \theta \circ l.$$

Axiom GA1 for group actions obviously holds, and GA2 is also satisfied as it amounts to nothing more, in this case, than the associativity property for the composition of mappings. Two labellings are different if they belong to different orbits of this action. So our problem amounts to counting the number of distinct orbits, and we can use Theorem 8.7 to do this.

The theorem is very easy to use in this case because all the stabilizers are trivial. The identity map ι_V is an automorphism which, as usual, fixes every element of our set, here $L(\mathcal{G})$, but no other automorphism fixes any labellings. For suppose θ is an automorphism of $\mathcal{G}$ and $\theta(v) \neq v$, for some $v \in V$. Since l is a bijection, there is some $i \in \{1, \ldots, k\}$ such that $l(i) = v$. Hence $\theta(l(i)) \neq l(i)$, and so $\theta \cdot l \neq l$.

Thus for each $l \in L(\mathcal{G})$, $S_l = \{\iota_V\}$, and hence $\#(S_l) = 1$. Hence

$$\sum_{l \in L(\mathcal{G})} \#(S_l) = \#(L(\mathcal{G})) = k!$$

and so, by Theorem 8.7, we have

Theorem 9.3
Let $\mathcal{G}$ be a graph with k vertices. The number of different labellings of $\mathcal{G}$ is

$$\frac{k!}{\#(\mathcal{A}(\mathcal{G}))},$$

where $\mathcal{A}(\mathcal{G})$ is the automorphism group of $\mathcal{G}$.

The formula given by Theorem 9.3 is very simple. When it comes to calculating the number of different labellings of particular graphs, you may think it is deceptively simple, since some thought is needed to determine the number of automorphisms of a graph. You may even think that it is just as straightforward to calculate directly the number of distinct labellings.

Whatever the difficulties with particular graphs, it is very easy to calculate the total number of different labellings of all the graphs with a particular number of vertices.

Theorem 9.4
There are $2^{k(k-1)/2}$ different labellings of graphs with k vertices.

Proof
The number of different labellings of graphs with k vertices is equal to the number of ways of choosing, for each two-element subset $\{i, j\}$ of $\{1, \ldots, k\}$ whether or not the vertices associated with the labels i and j are connected or not.

Now $\{1, \ldots, k\}$ has $C(k, 2) = k(k-1)/2$ subsets containing two elements and for each of these subsets we have two choices, either to connect the vertices with those labels or not. Hence there are $2^{k(k-1)/2}$ ways to choose a labelled graph with k vertices.

Exercises

9.2.1. List all the different graphs with one, two, three and four vertices. For each of these graphs calculate how many labellings it has.

9.2.2. Give the proof of Lemma 9.2.

10

Counting patterns

10.1 BURNSIDE'S THEOREM

In this section we show how the formula of Theorem 8.7 can be rewritten in a form which is much easier to use. Then in section 10.2 we show how this improved version of the theorem can be used to solve many counting problems.

Theorem 8.7 tells us that if the group G acts on a set X, then the number of distinct orbits is

$$\frac{1}{\#(G)} \sum_{x \in X} \#(S_x).$$

When X was the set of colourings of a 2×2 chessboard, where $\#(X) = 16$, it was easy to calculate the sum in this formula. However, for an 8×8 chessboard, where $\#(X) > 10^{19}$, it is completely impractical to do this. We can, however, exploit the fact that although the set of colourings of an 8×8 chessboard is very large, the group G of symmetries remains the same as in the case of a 2×2 chessboard. It is still the eight-element symmetry group of the square. Thus if we can replace the sum, given in Theorem 8.7, over X by a sum over G, the problem of evaluating the sum becomes tractable.

We set about this in the following way. We consider the table shown below, where we have rows corresponding to the elements G and columns corresponding to the elements of X.

We put a tick in the g_i row and the x_j column if and only if

$$g_i \cdot x_j = x_j,$$

that is, if and only if $g_i \in S_{x_j}$. Thus the number of ticks in the x_j column is $\#(S_{x_j})$. Hence the total number of ticks in the table is $\Sigma_{x \in X} \#(S_x)$.

We can count the number of ticks in this table in another way. Instead of adding up the column totals, we can add up the row totals.

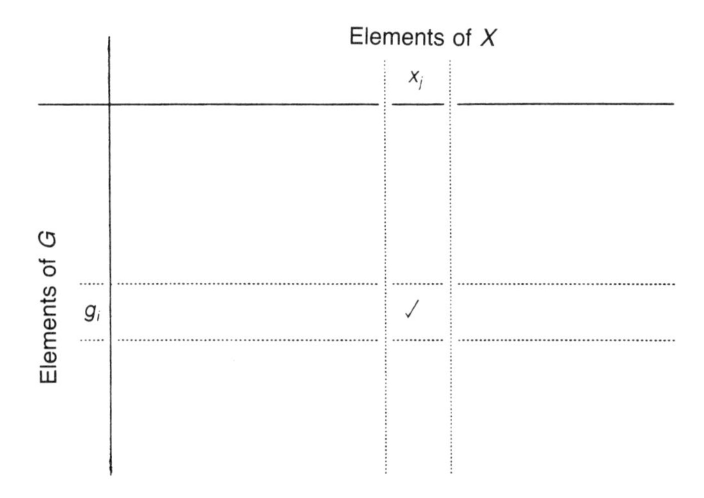

The number of ticks in the g_i row is the number of elements in the set

$$\{x \in X : g_i \cdot x = x\}.$$

This is the set of those elements of X which are fixed by the group element g_i. We call it the **fix** of g_i, written $\text{Fix}(g_i)$. The sum of the row totals is thus $\Sigma_{g \in G} \#(\text{Fix}(g))$. The key idea that we now use is very simple: the sum of the row totals must be the same as the sum of the column totals, as each sum gives the total number of ticks in the table. The surprising thing is that this very simple idea turns out to be so useful. The formula which expresses this idea is

$$\sum_{g \in G} \#(\text{Fix}(g)) = \sum_{x \in X} \#(S_x). \tag{10.1}$$

Thus we can replace the sum $\Sigma_{x \in X} \#(S_x)$ that occurs in Theorem 8.7 by the sum on the left-hand side of equation (10.1), and thus deduce the following important theorem.

Theorem 10.1 *(Burnside's Theorem)*
If G is a finite group which acts on the set X, then the number of distinct orbits is

$$\frac{1}{\#(G)} \sum_{g \in G} \#(\text{Fix}(g)).$$

Although the formula in Burnside's Theorem is only a variation of the formula in Theorem 8.7, it is very much more useful. As we illustrate in the examples in the next section, we can use it to count the number of different patterns in many different cases. In particular, we can use Burnside's Theorem to solve the question about chessboards with which we began Chapter 8.

William Burnside, after whom this theorem is named, lived from 1852 to 1927. He received his mathematical education in Cambridge, and in 1885 became Professor of Mathematics at the Royal Naval College at Greenwich. He was originally thought of as an applied mathematician, but his first published paper concerned elliptic functions, and later in his life his interests turned to group theory. He published the first edition of his influential book on group theory, the first such book to be published in English, in 1897, and a second edition appeared in 1911.*

However, it is a historical accident that this theorem has come to have Burnside's name attached to it, since it was first proved by Georg Frobenius in 1887. Burnside gives the theorem in his text. In the first edition, he attributed it to Frobenius, but this got missed out in the more widely read second edition. This seems to be the source of the confusion.† I have decided to stick to the common, though historically inaccurate, name for this result.

10.2 APPLICATIONS OF BURNSIDE'S THEOREM

We return now to the question of the number of different patterns that can be obtained by colouring the 64 squares of an 8×8 chessboard using the colours black and white.

Our underlying set, X, is the set of all the 2^{64} ways of colouring the squares of the chessboard using these two colours, and the group, G, is the group of the eight symmetries of the square. It follows from Burnside's Theorem that, in order to answer our question, we need to calculate $\#(\text{Fix}(g))$ for each element g of G.

Recall that we are using the following notation for the symmetries of the square:

e the identity
a rotation through $\pi/2$ clockwise.
b rotation through π clockwise.
c rotation through $3\pi/2$ clockwise.

p reflection in the horizontal axis. 
q reflection in the vertical axis. 
r reflection in the diagonal axis shown 
s reflection in the diagonal axis shown

*William Burnside, *Theory of Groups of Finite Order*, Cambridge University Press, 1897 (2nd edn 1911).

†For some historical remarks about the theorem see P. A. Neumann, 'A lemma that is not Burnside's', *Math. Scientist*, 4 (1979), pp. 133–41; and E. M. Wright, 'Burnside's Lemma: A historical note', *Journal of Combinatorial Theory B* 25 (1981), pp. 89–90.

Clearly the identity, e, fixes every colouring so $\#(\mathrm{Fix}(e)) = 2^{64}$. Now let us consider $\mathrm{Fix}(a)$. What property must a colouring have to be fixed by a, that is, to look the same after the chessboard has been rotated through an angle $\pi/2$?

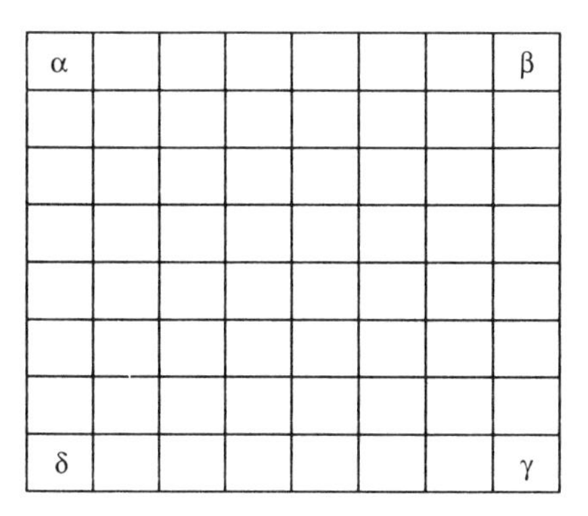

The effect of the symmetry a is to move the square labelled α to the position of the square labelled β, β to γ, γ to δ, and δ back to α. Therefore, if the chessboard is to look the same after the rotation as it did before, all the four squares in this cycle must have the same colour. Every other square belongs to a similar cycle of four squares, and if the chessboard is to look the same after it has been rotated through $\pi/2$, all the four squares in each of these cycles must have the same colour.

Thus, corresponding to the symmetry, a, we have a permutation, say, π_a, of the 64 squares which make up the chessboard. The cycle type of π_a is x_4^{16}. That is, π_a is made up of 16 cycles each of length 4. A colouring looks the same after the symmetry a has been applied to the chessboard, if and only if all the squares in each of these 16 cycles are the same colour. Thus in choosing a colouring which is fixed by the symmetry a, we are free to assign a colour to each of these 16 cycles; the colour which all the squares in that cycle must have.

Since there are two possible colours for each cycle, and 16 cycles, it follows that we have 2^{16} choices altogether. Thus

$$\#(\mathrm{Fix}(a)) = 2^{16}.$$

Another way of viewing this result is that once we have decided how to colour the 16 squares in the top left-hand corner of the chessboard, then the colours of all the remaining squares are determined if the board is to look the same after the rotation through a quarter turn. So the number of colourings in $\mathrm{Fix}(a)$ is the same as the number of ways of colouring the 16 squares in the top left-hand corner, that is 2^{16}.

It is important to understand the idea which lies behind this calculation of $\#(\mathrm{Fix}(a))$. You are therefore encouraged to pause in your reading and to calculate #(Fix(g)) for every other element g of the symmetry group of a square. Then check whether your answers agree with the numbers in the table below.

You should have tackled this problem by considering the permutation π_g of the 64 squares of the chessboard, corresponding to each symmetry g. All the squares in each cycle of π_g must have the same colour if the chessboard is to look the same after the symmetry g has been carried out. Thus, if π_g has n cycles, then $\#(\text{Fix}(g)) = 2^n$. In this way you should obtain the following values:

g	e	a	b	c	p	q	r	s
#(Fix(g))	2^{64}	2^{16}	2^{32}	2^{16}	2^{32}	2^{32}	2^{36}	2^{36}

We can now use Burnside's Theorem to calculate the total number of different patterns. All we have to do is to add up the values of #(Fix(g)) and then divide the total by 8, the number of symmetries in the group. Thus we can deduce that the total number of different patterns is

$$\tfrac{1}{8}(2^{64} + 2(2^{36}) + 3(2^{32}) + 2(2^{16})) = 2\,305\,843\,028\,004\,192\,256.$$

Rather a large number! You will note that the dominant term in the above expression is 2^{64}, corresponding to the identity symmetry. Hence dividing the total number of colourings in X by the number of elements in G gives a good approximation to the correct value.

It is easy to see that if we have more than two colours, all we need do is to replace the 2 in the formula above by the number of colours we are using. Thus the number of different patterns that can be obtained by colouring the squares of an 8×8 chessboard using c different colours is

$$\tfrac{1}{8}(c^{64} + 2c^{36} + 3c^{32} + 2c^{16}).$$

We now look at two more examples where we can use Burnside's Theorem. You are encouraged to try them for yourself before reading the solution.

Problem 1

In how many ways can the vertices of a regular hexagon be coloured using c different colours?

Solution

This is exactly the same problem as that of the number of patterns that can be formed by arranging six beads in a ring, using beads of c colours.

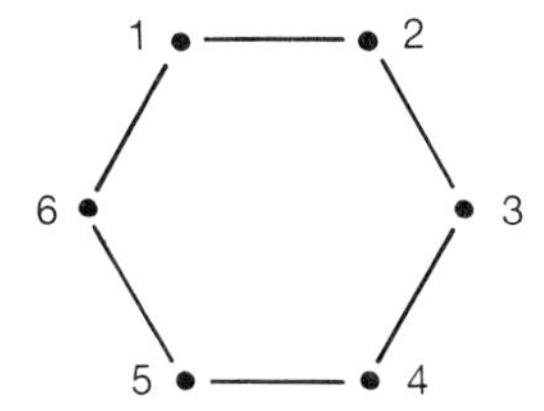

The relevant set, X, here is the set of all ways of choosing colours for the six vertices. Since there are c colours to choose from, $\#(X) = c^6$. The relevant group is the group, G, of symmetries of a regular hexagon. You will recall from Chapter 7 that there are 12 of these symmetries as follows:

e	the identity,
r_i	rotation through $2i\pi/6$ clockwise, for $i = 1, 2, \ldots, 5$.
s_i	reflection in an axis joining opposite vertices, for $i = 1, 2, 3$.
t_i	reflection in an axis joining the midpoints of opposite edges, for $i = 1, 2, 3$.

In order to work how many different patterns there are we need to work out how many colourings are fixed by each of these symmetries. We can work this out by considering the permutations of the vertices corresponding to each of these symmetries. As with the colourings of a chessboard, a colouring is fixed by a symmetry if and only if all the vertices in each of the cycles of this permutation are assigned the same colour. Hence, if the permutation is made up of k different cycles, the number of colourings fixed by the symmetry is c^k.

So consider, for example, the rotation through $2\pi/3$, that is, a rotation of one-third of a turn clockwise, which we have denoted by r_2. This symmetry corresponds to the permutation $(1\,3\,5)(2\,4\,6)$ of the vertices. Since this permutation is made up of two cycles, $\#(\text{Fix}(r_2)) = c^2$. Repeating this for all the symmetries in turn we get the following values:

g	e	r_1	r_2	r_3	r_4	r_5	s_i	t_i
$\#(\text{Fix}(g))$	c^6	c^1	c^2	c^3	c^2	c^1	c^4	c^3

Hence it follows from Burnside's Theorem that the number of different patterns is

$$\tfrac{1}{12}(c^6 + 3c^4 + 4c^3 + 2c^2 + 2c).$$

Notice that since the number of patterns must be a whole number, we have shown that for every positive integer c the polynomial

$$c^6 + 3c^4 + 4c^3 + 2c^2 + 2c$$

is always divisible by 12.

Problem 2

How many different ways are there to colour the faces of a regular tetrahedron using c colours? (Two colourings should be regarded as the same if one can be obtained from the other by a rotational symmetry.)

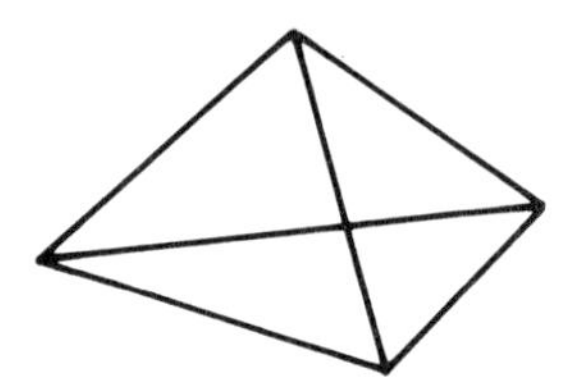

Solution

The comment in brackets indicates that the group we are going to use here is the group of rotational symmetries of a regular tetrahedron. We have chosen to work just with the rotational symmetries because these can be carried out physically, and so they are easy to picture. You are strongly encouraged to make your own model of a regular tetrahedron so that you can carry out these symmetries for yourself.

A tetrahedron has four faces and so there are c^4 different colourings using c colours. We consider the different rotational symmetries of a tetrahedron and we work out the number of colourings fixed by each of them. We do this by considering the cycle type of the associated permutations of the faces of the tetrahedron using the same method as with the earlier problems in this section.

We saw in the solution to Exercise 7.3.3 that the rotational symmetries of a tetrahedron fall into three categories. (Readers experienced in group theory will notice that these categories are the *conjugacy classes* of the group. Conjugate permutations have the same cycle type.) The first is the identity element e. The corresponding permutation of the faces has cycle type x_1^4, and so $\#(\mathrm{Fix}(e)) = c^4$.

The second is made up of rotations through one-third of a turn, clockwise and anti-clockwise (i.e. rotations through $2\pi/3$) about axes joining a vertex to the midpoint of the opposite face. Since there are four of these axes and two rotations about each of them, there are eight symmetries of this sort. The corresponding permutations of the faces all have cycle type x_1x_3 and hence each of them fixes c^2 colourings.

The third consists of rotations through one-half of a turn (i.e. through π) about axes joining the midpoints of opposite edges. There are three axes joining the midpoints of opposite edges and hence three symmetries of this type. The cycle type of the corresponding permutations of the faces is x_2^2, and hence each of these symmetries fixes c^2 colourings.

Thus it turns out that $\#(\text{Fix}(e)) = c^4$, and that for every other symmetry g, $\#(\text{Fix}(g)) = c^2$. Hence, by Burnside's Theorem, the number of different patterns is

$$\tfrac{1}{12}(c^4 + 11c^2).$$

Exercises

10.2.1. How many different patterns can be formed by colouring the edges of a triangle using 4 colours? (Consider just the rotational symmetries. By coincidence, a puzzle involving these different colourings appeared in the *Sunday Correspondent*, under the title, 'Infernal Triangles', as the final draft of this book was being written.)

10.2.2. How many different patterns can be formed by colouring the vertices of a regular octagon using c colours?

10.2.3. How many different patterns can be formed by colouring the squares of a 5×5 chessboard using 3 colours?

10.2.4. How many different patterns can be formed by colouring the small squares in the following figure using the colours red, white and blue?

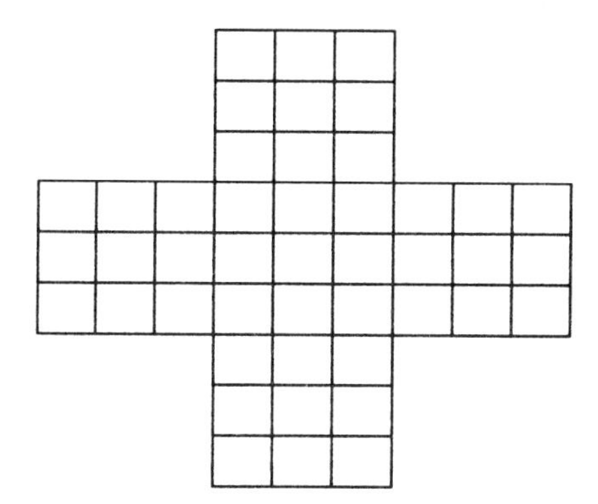

10.2.5. In how many different ways can you colour the faces of a cube using the colours red, white and blue? (Two colourings should be regarded as the same if one can be obtained from the other by a rotational symmetry of the cube. You are encouraged to find a cube, or make yourself one, so that you can work out all its rotational symmetries.)

10.2.6. In how many different ways can the numbers 1 to 6 be placed on the faces of a cube? (Again, use the group of rotational symmetries of the cube.)

10.2.7. There are altogether five Platonic regular solids. In addition to the tetrahedron and cube, which we have already considered, there are the regular octahedron, dodecahedron and icosahedron. For each of these three latter regular solids, calculate the number of patterns that can be obtained by colouring their faces using c different colours. (Again, consider only the rotational symmetries.)

A note on models

As I have already stressed, when tackling problems of this kind it helps to have a model of the relevant geometrical figure, to help you to work out the symmetry group. Here are some suggestions for making these models.

The book *Mathematical Models* by H. M. Cundy and A. P. Rollett, gives full instructions for making models of all the regular solids, (and also other shapes whose colourings you might like also to consider). *Make Shapes (Series No. 1): 19 mathematical models to cut out, glue and decorate,* as the title suggests, contains card shapes which can be cut out and glued to form the five Platonic regular solids, and 14 other shapes. Both these books are published by Tarquin Publications, Stradbroke, Diss, Norfolk.

An easier, but more expensive, way to make regular solids is to use the plastic POLYDRON™ shapes made by Polydron UK Ltd, and available from good toy shops.

I should be very interested to hear from any reader who knows of other easy ways to obtain models of the regular solids.

11

Pólya's theorem

11.1 COLOURINGS AND GROUP ACTIONS

Burnside's Theorem enables us to count the total number of patterns quite easily in the types of case we have dealt with in Chapter 10. However, it is not so easy to use when it comes to more delicate problems such as how many different patterns can be obtained by colouring the squares of a chessboard so that there are 32 black squares and 32 white squares, or how many different positions can arise in a game of noughts and crosses.

The game of noughts and crosses (also called tic-tac-toe) is played on this sort of grid:

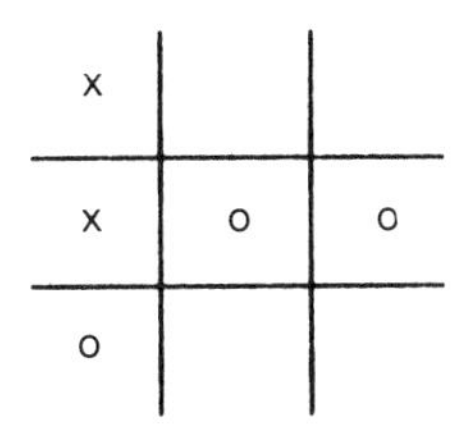

Each of the nine spaces in the grid can be occupied by a nought or a cross or is blank. Thus positions in the game correspond to the colourings of a 3×3 chessboard using three colours. However, the rules of the game mean that certain positions cannot occur. For example, if 'noughts' goes first the number of noughts must always equal the number of crosses, or be one more than the number of crosses. Hence the problem of counting the number of different positions is rather more complicated than the problem of simply counting the number of different colourings of the chessboard.

We are going to develop the theory which enables us to handle problems of this kind. First, we need to look a little more carefully at what we mean by a **colouring** and at how the symmetry group of the underlying figure acts on the set of colourings.

It is again helpful to consider a particular example first. We will therefore return to our favourite example, that of a 2×2 chessboard.

α	β
δ	γ

We can regard a colouring, using black and white, as a mapping from the set of squares into which the chessboard is divided to the set of colours. Thus if we let $\{\alpha, \beta, \gamma, \delta)$ be the set of squares, as shown, then each colouring is a mapping

$$f : \{\alpha, \beta, \gamma, \delta\} \to \{\text{black, white}\}.$$

The 16 different colourings then correspond to the 16 different mappings. Using the same numbering of the colourings as we used in Figure 8.1, we let f_i be the mapping corresponding to the colouring Ci. For example C6 is the colouring which corresponds to the mapping f_6 given by

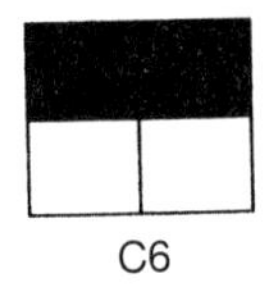

$$f_6(\alpha) = f_6(\beta) = \text{black}$$

$$f_6(\gamma) = f_6(\delta) = \text{white}.$$

Corresponding to each symmetry, g, of the square there is a permutation, π_g, of the set of squares $\{\alpha, \beta, \gamma, \delta\}$. For example, corresponding to the quarter turn we have the permutation

$$\pi_a = (\alpha\beta\gamma\delta).$$

We can take as our group the group of all these permutations of the set $\{\alpha, \beta, \gamma, \delta\}$. How does it act on the set of colourings? We can see how the action is defined if we consider what happens when the permutation π_a acts on the colouring f_6.

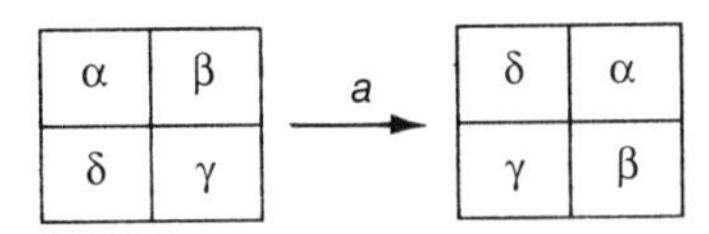

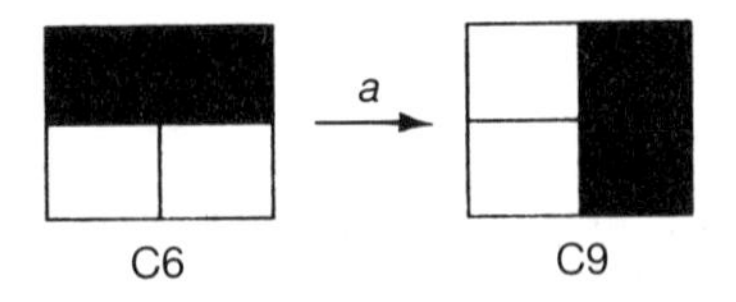

Thus $\pi_a \cdot f_6 = f_9$, as the rotation a takes us from the colouring f_6 to the colouring f_9. Looking at this more closely, we see that $\pi_a \cdot f_6$ assigns to square α the colour white, because this is the colour which f_6 assigns to the square δ, and π_a moves δ to α, or, in other words $\pi_a^{-1}(\alpha) = \delta$. Thus, in general the colour which $\pi_a \cdot f_6$ assigns to a square χ, is the same as the colour which f_6 assigns to the square which π_a moves to the position of χ. In symbols, for each square χ,

$$\pi_a \cdot f_6(\chi) = f_6(\pi_a^{-1}(\chi)).$$

It follows that the action of π_a on f_6 is simply to compose it with π_a^{-1}, that is,

$$\pi_a \cdot f_6 = f_6 \circ \pi_a^{-1}$$

This formula applies, in general, to the action of the permutations of the set $\{\alpha, \beta, \gamma, \delta\}$ on the colourings of the chessboard, and, indeed, to other situations of a similar kind.

We can now describe the general situation as follows. Necessarily this description is rather abstract, but the reason for it will be clearer if you keep in mind the particular example that we have just discussed in some detail.

Let C, D be two different sets. We let X be the set of all mappings from D to C. ('D' for 'domain' and 'C' for 'codomain', but you can also think of 'C' as standing for the set of colours which are assigned to elements of D.) We call the mappings which make up the elements of X the **colourings** of D.

Now suppose that G is some group of permutations of D. Thus G will be a subgroup of the group $S(D)$ of all the permutations of D. We define an action of G on the set of colourings, X, by

$$\text{for } \pi \in G \text{ and } f \in X,$$
$$\pi \cdot f = f \circ \pi^{-1}. \tag{11.1}$$

We need to check that this does define a group action.

Lemma 11.1

Formula (11.1) does define a group action.

Proof

Let ι be the identity element of G. Then ι is the identity map $\iota: D \to D$ and for each $f \in X$,

$$\iota \cdot f = f \circ \iota^{-1} = f \circ \iota = f,$$

and thus GA1 holds.

Now let π_1, π_2 be two permutations from G, and suppose $f \in X$. Then

$$\begin{aligned}\pi_2\cdot(\pi_1\cdot f) &= \pi_2\cdot(f \circ \pi_1^{-1})\\ &= (f \circ \pi_1^{-1}) \circ \pi_2^{-1}\\ &= f \circ (\pi_1^{-1} \circ \pi_2^{-1})\\ &= f \circ (\pi_2 \circ \pi_1)^{-1}\\ &= (\pi_2 \circ \pi_1)\cdot f,\end{aligned}$$

and this shows that GA2 also holds, thus completing the proof.

We have already seen (see Exercise 8.2.2) that the action of each element of G is to permute X. Thus for each $\pi \in G$, the mapping $\Phi : X \mapsto X$ defined, for $f \in X$, by

$$\Phi(f) = f \circ \pi^{-1},$$

is a permutation of X.

11.2 PATTERN INVENTORIES

In this section we are going to introduce an algebraic method for describing colourings. The ultimate outcome will be an algebraic expression telling us how many patterns of each kind there are.

Colouring a 2×2 chessboard with the colours black and white involves choosing for each of the four squares one of these two colours. This should remind us of algebraic expansions. If we expand the algebraic expression

$$(b + w)^4 = (b + w)(b + w)(b + w)(b + w),$$

then we obtain each term by choosing from each of the four brackets either b or w. Hence the terms in the expansion of this expression correspond to the different colourings of the chessboard. Now,

$$(b + w)^4 = b^4 + 4b^3w + 6b^2w^2 + 4bw^3 + w^4.$$

The term b^4 comes by choosing b from each bracket. This corresponds to colouring each of the four squares black. Likewise the term b^3w arises by choosing b from three of the brackets and w from the fourth bracket. This corresponds to colouring three of the squares black and the remaining square white. Thus the coefficient of b^3w indicates that there are four colourings of this kind. Likewise the $6b^2w^2$ in the expansion tells us that there are 6 colourings with two black and two white squares, and so on.

This algebraic expression thus tells us how many colourings there are with a given number of black squares and a given number of white squares. However it does not tell us anything about the number of different patterns. We are aiming towards a theorem which does indeed provide a similar expression telling us how many different patterns there are with a given number of squares of each colour.

First, we need to be more precise about the sort of expression we are looking for. Again, the abstract general description will be clearer if you keep in mind a particular example like the one above.

In this example the symbols b and w corresponded to the colours black and white. In general by a **weight function** on the set C we mean a mapping ω which assigns to each $c \in C$ an algebraic symbol $\omega(c)$.

The sum $\Sigma_{c \in C}\ \omega(c)$ is called the **store enumerator**. For each colouring $f \in X$, the **weight** of f, written $W(f)$ is the algebraic expression

$$\prod_{d \in D} \omega(f(d)).$$

If S is a subset of X the **inventory** of S is the expression

$$\sum_{f \in S} W(f).$$

These definitions will be made clearer by remembering our example involving the 2×2 chessboard. In this example the weight function is the mapping ω such that

$$\omega(\text{black}) = b$$
$$\omega(\text{white}) = w.$$

The colouring f_6 is such that

$$f_6(\alpha) = \text{black},$$
$$f_6(\beta) = \text{black},$$
$$f_6(\gamma) = \text{white},$$
$$f_6(\delta) = \text{white}.$$

Hence

$$\begin{aligned} W(f_6) &= \prod_{d \in D} \omega(f_6(d)) = \omega(f_6(\alpha))\omega(f_6(\beta))\omega(f_6(\gamma))\omega(f_6(\delta)) \\ &= \omega(\text{black})\omega(\text{black})\omega(\text{white})\omega(\text{white}) \\ &= bbww \\ &= b^2 w^2. \end{aligned}$$

One of the uses of inventories will be explained by the following problems.

Problem 1

Suppose we have a triangle and three sorts of paints: light red paint, dark red paint, and yellow paint. How many different colourings are there of the three edges of the triangle?

Solution

Here D is the set of edges of the triangle and C is the set of paints. Let ω be the weight function which assigns to these paints the weights r_1, r_2 and y respectively. Thus the store enumerator is

$$r_1 + r_2 + y,$$

and the inventory of the set of all the colourings is

$$(r_1 + r_2 + y)^3 = r_1^3 + r_2^3 + y^3 + 3r_1r_2^2 + 3r_1^2r_2 + 3r_1y^2 + 3r_1^2y + 3r_2y^2 + 3r_2^2y + 6r_1r_2y$$

which tells us, for example, that there are three colourings in which dark red paint is used for two of the edges and yellow paint for one edge.

If we give both of the red paints the weight r, the store enumerator becomes $r + r + y$, which we can write as $2r + y$, and hence the inventory is

$$(2r + y)^3 = 8r^3 + 12r^2y + 6ry^2 + y^3,$$

from which we can deduce that, for example, there are eight colourings of the triangle in which all three of the edges are red.

Problem 2

Ten mathematics students each have to do a project. The choice of projects for each student is from

six projects in pure mathematics,
four projects in applied mathematics,
two projects in statistics, and
two projects in computing.

Because of staffing shortages, the projects must be chosen so that

four students do pure mathematics projects,
three students do applied mathematics projects,
two students do statistical projects, and
one student does a computer project.

Two or more students can choose to undertake the same project, subject to the constraints above. In how many different ways can the students choose their projects?

Solution

Here D is the set of ten students, C is the set of 14 projects. We assign the weights p, a, s, c to the projects in pure mathematics, applied mathematics, statistics and computing, respectively.

The store enumerator is thus

$$(6p + 4a + 2a + 2s)$$

and the inventory of all the possible choices by the students, ignoring the constraints, is

$$(6p + 4a + 2s + 2c)^{10}.$$

The constraint on the choice of projects implies that the relevant term in the expansion is the one involving $p^4a^3s^2c$. By the Multinomial Theorem, (exercise 1.3.8) the relevant term is

$$\frac{10!}{4!3!2!1!}(6p)^4(4a)^3(2s)^2 2c = 8\,360\,755\,200p^4a^3s^2c,$$

and hence there are 8 360 755 200 ways in which the projects can be chosen.

The problem of most interest to us is the inventory of a set of colourings which contains exactly one colouring of each pattern. We call this the **pattern inventory**. In the next section we begin the task of evaluating this.

Exercises

11.2.1. Six English literature students have to choose a novel about which to write a dissertation. The novel must be by Jane Austen, Charles Dickens or Anthony Trollope. Two students may choose the same novel but, again because of staffing problems in allotting supervisors—English Departments have also suffered from the effect of cuts in universities—the novels must be chosen so that altogether there are two by each author. Given that Jane Austen published six novels, Charles Dickens 14 and Anthony Trollope 48, in how many ways can the students make their choices?

11.2.2. Write down the pattern inventory for the colourings of a 2×2 chessboard using two colours.

11.3 THE CYCLE INDEX OF A GROUP

You will remember from section 7.6 that the **cycle type** of a permutation is the algebraic expression

$$x_{l_1}^{k_1}x_{l_2}^{k_2} \ldots x_{l_s}^{k_s}$$

where, in the disjoint cycle decomposition of the permutation, there are, for $1 \leqslant i \leqslant s$, k_i cycles of length l_i. We will use the notation $\mathrm{ct}(\pi)$ to stand for the cycle type of the permutation π.

Now comes a key definition.

If G is a group of permutations, the **cycle index** of G, denoted C_G, is defined to be the polynomial

$$\frac{1}{\#(G)} \sum_{\pi \in G} \mathrm{ct}(\pi).$$

Notice that the cycle index of a group is a polynomial in unknowns x_1, $x_2, \ldots$, and that the sum of the coefficients in this polynomial is always equal to 1. This is because there are $\#(G)$ terms in the sum, which is then divided by $\#(G)$.

We illustrate this idea by showing how to calculate the cycle index of the groups S_n for $n = 1, 2, 3, \ldots$.

We start with the trivial case of the group S_1 of all permutations of one object. There is just one permutation in S_1, namely the identity, which has cycle type x_1. Hence

$$C_{S_1} = x_1.$$

Next consider the group S_2. There are two permutations in S_2: the identity (1)(2) which has cycle type x_1^2 and the permutation (1 2) which has cycle type x_2. Hence the cycle index of S_2 is given by

$$C_{S_2} = \tfrac{1}{2}(x_1^2 + x_2).$$

We list the six permutations in S_3 and their cycle types:

Permutation	*Cycle Type*
(1)(2)(3)	x_1^3
(1 2), (1 3), (2 3)	$x_1 x_2$
(1 2 3), (1 3 2)	x_3

Hence the cycle index of S_3 is

$$C_{S_3} = \tfrac{1}{6}(x_1^3 + 3x_1x_2 + 2x_3).$$

As n becomes larger it becomes more and more cumbersome to list all the elements of S_n. We have already seen (in Chapter 7) that the different cycle types in S_n correspond to the different partitions of n. So for S_4 we will simply list the partitions of 4, the corresponding cycle types and the number of permutations in S_4 of this type.

Partition	*Cycle type*	*No. of permutations*
4	x_4	6
3 + 1	x_1x_3	8
2 + 2	x_2^2	3
2 + 1 + 1	$x_1^2x_2$	6
1 + 1 + 1 + 1	x_1^4	1

It follows that

$$C_{S_4} = \tfrac{1}{24}(x_1^4 + 6x_1^2x_2 + 3x_2^2 + 8x_1x_3 + 6x_4).$$

I have not explained how I calculated the number of permutations of each cycle type. Before I do this, try one for yourself. Calculate the cycle index of the group S_5. (This is repeated as Exercise 11.2.1, so a solution can be found among the solutions for these Exercises.)

We now calculate the number of permutations in S_n of cycle type

$$x_{l_1}^{k_1}x_{l_2}^{k_2}\ldots x_{l_s}^{k_s}$$

corresponding to the partition

$$k_1l_1 + k_2l_2 + \ldots + k_sl_s = n \tag{11.2}$$

of the number n. We count the number of permutations of this cycle type by counting the number of ways we can arrange the numbers from the set $\{1, 2, \ldots, n\}$ into a permutation of this type.

Suppose that we first choose the numbers to make up the k_1 cycles of length l_1. The numbers to make up the first of these cycles can be chosen in $C(n, l_1)$ ways. l_1 numbers can be arranged to make $(l_1 - 1)!$ different cycles of length l_1. (We can arrange l_1 numbers in order in $l_1!$ ways, but each cycle of length l_1 can be written in l_1 different ways, depending on which number we put first. So there are $l_1!/l_1 = (l_1 - 1)!$ different cycles of length l_1 made up from a given set of l_1 numbers.) We then have left $(n - l_1)$ numbers from which to make up the next cycle of length l_1, and so these numbers can be chosen in $C(n - l_1, l_1)$ ways, and again they can be arranged to form $(l_1 - 1)!$ different cycles of length l_1. And so on until we have chosen k_1 cycles of length l_1. Now the order in which we arrange these k_1 cycles of length l_1 is irrelevant. Hence the total number of ways we can choose these cycles is

$$\frac{1}{k_1!}\big(C(n, l_1)(l_1 - 1)!C(n - l_1, l_1)(l_1 - 1)! \ldots C(n - (k_1 - 1)l_1, l_1)(l_1 - 1)!\big)$$
$$= \frac{((l_1 - 1)!)^{k_1}}{k_1!}\big(C(n, l_1)C(n - l_1, l_1)C(n - 2l_1) \ldots C(n - (k_1 - 1)l_1, l_1)\big)$$
$$= \frac{((l_1 - 1)!)^{k_1}}{k_1}\left(\frac{n!}{l_1!(n - l_1)!} \times \frac{(n - l_1)!}{l_1!(n - 2l_1)!} \times \ldots \times \frac{(n - (k_1 - 1)l_1)!}{l_1!(n - k_1l_1)!}\right)$$
$$= \frac{n!}{l_1^{k_1}k_1!(n - k_1l_1)!}, \qquad (11.2)$$

after a lot of cancelling.

Having chosen k_1 cycles of length l_1, we are left with $(n - k_1l_1)$ numbers from which to choose k_2 cycles of length l_2. The same reasoning that we used to obtain formula (11.2) above, shows that we can do this in

$$\frac{(n - k_1l_1)!}{l_2^{k_2}k_2!\,(n - k_1l_1 - k_2l_2)!}$$

ways, and so on.

It follows that the total number of permutations in S_n of cycle type $x_{l_1}^{k_1}x_{l_2}^{k_2}\ldots x_{l_s}^{k_s}$ is

$$\frac{n!}{l_1^{k_1}k_1!(n - k_1l_1)!} \times \frac{(n - k_1l_1)!}{l_2^{k_2}k_2!(n - k_1l_1 - k_2l_2)!} \times \ldots \times \frac{(k_sl_s)!}{l_s^{k_s}k_s!0!}$$

(The numerator in the last term in this product arises as follows. According to the general formula we are using it should be $(n - k_1l_1 - k_2l_2 - \ldots - k_{s-1}l_{s-1})!$, but it follows from equation (11.1) above that the term in the bracket is equal to k_sl_s.)

All the terms in the numerator of the last displayed expression above, with the exception of the term $n!$, cancel with terms in the denominator, and so we are left with

$$\frac{n!}{l_1^{k_1}k_1!l_2^{k_2}k_2! \ldots l_s^{k_s}k_s!}$$

as the number of permutations of cycle type $x_{l_1}^{k_1}x_{l_2}^{k_2}\ldots x_{l_2}^{k_2}$ in S_n.

The cycle index of S_n is, then, the sum of all the terms of the above form multiplied by $x_{l_1}^{k_1}\ldots x_{l_s}^{k_s}$ for the combinations of positive integers s, $k_1, \ldots, k_s$, and $l_1, \ldots, l_s$ satisfying equation (11.1), divided by $n!$, which is the number of elements in S_n.

Exercises

11.3.1. Calculate the cycle index of the groups S_5 and S_6.

11.3.2. Work out the cycle index of each of the following groups of permutations:

(a) The subgroup of S_4 corresponding to the symmetries of a square.

(b) The subgroup of S_6 corresponding to the symmetries of a regular hexagon.

(c) The subgroup of S_6 corresponding to the rotational symmetries of the cube.

(In cases (a) and (b) it makes no difference whether you take the permutations to be the permutations of the vertices, or of the edges, corresponding to the symmetries of the figure. In case (c) you should take the group of permutations to be those of the faces of the cubes corresponding to the rotational symmetries.)

11.4 PÓLYA'S THEOREM: STATEMENT AND EXAMPLES

We are now ready to state the theorem which tells us how to calculate pattern inventories. Having stated this theorem, we illustrate it with some examples. The proof of the theorem is postponed until the next section.

This theorem gets its name from the Hungarian mathematician George Pólya (born 13 December 1887, died 7 September 1985), who published it in 1937. An English translation of this paper has recently been published.* In addition to the translation of Pólya's paper, this book includes a survey by R. C. Read of research in this area since the paper was published. It turns out that Pólya's work was to some extent anticipated in a paper published by J. H. Redfield in 1927, but which went unnoticed for many years.

Theorem 11.2 *Pólya's Theorem*

Let X be the set of all mappings from the set D to the set C, and let ω be a weight function on C. Let G be a group of permutations of D which acts on X in the standard way. If the cycle index of G is

$$C_G(x_1, x_2, x_3, \ldots)$$

then the pattern inventory is

$$C_G\left(\sum_{c\in C}\omega(c), \sum_{c\in C}\omega(c)^2, \sum_{c\in C}\omega(c)^3, \ldots\right).$$

*G. Pólya and R. C. Read, *Combinatorial Enumeration of Groups, Graphs, and Chemical Compounds*, Springer-Verlag, 1987.

This formula means that we obtain the pattern inventory by taking the cycle index of the group G, and replacing each occurrence of x_1 by the sum of the weights, each occurrence of x_2 by the sum of the squares of the weights, each occurrence of x_3 by the sum of the cubes of the weights, and so on. We illustrate the use of this formula by several examples before giving a proof of the theorem.

Example 1

We return to our standard example of 2×2 chessboards. Of course, we already know about the patterns in this case, but it will be helpful to start with a simple example, where the calculation is very straightforward.

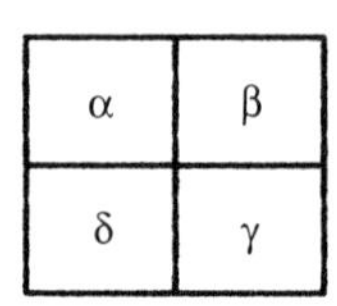

Here $D = \{\alpha, \beta, \gamma, \delta\}$. $C = \{\text{black, white}\}$ and our weight function, ω, is given by

$$\omega(\text{black}) = b, \qquad \omega(\text{white}) = w.$$

The group G is the group of permutations of D corresponding to the eight symmetries of the square. Thus

$$G = \{e, (\alpha\beta\gamma\delta), (\alpha\gamma)(\beta\delta), (\alpha\delta\gamma\beta), (\alpha\delta)(\beta\gamma), (\alpha\beta)(\gamma\delta), (\beta\delta), (\alpha\gamma)\}$$

and thus the cycle index of G is given by

$$C_G(x_1, x_2, x_4) = \tfrac{1}{8}(x_1^4 + 2x_1^2x_2 + 3x_2^2 + 2x_4). \qquad (11.3)$$

By Pólya's Theorem we now obtain the pattern inventory by substituting $b + w$ for x_1, $b^2 + w^2$ for x_2 and $b^4 + w^4$ for x_4 in equation (11.3). Thus the pattern inventory is

$$\tfrac{1}{8}((b + w)^4 + 2(b + w)^2(b^2 + w^2) + 3(b^2 + w^2)^2 + 2(b^4 + w^4))$$
$$= b^4 + b^3w + 2b^2w^2 + bw^3 + w^4. \qquad (11.4)$$

This pattern inventory, expression (11.4), tells us that when we colour the 2×2 chessboard using the two colours black and white, we get

one pattern with 4 black squares
one pattern with 3 black and 1 white square,
two patterns with 2 black and 2 white squares,
one pattern with 1 black and 3 white squares, and
one pattern with 4 white squares.

Example 2

Next we consider the computationally more complicated example of the colourings of a 3×3 chessboard using three colours, which we will take to be red, white and blue.

1	2	3
4	5	6
7	8	9

Here D is the set of the nine squares of the chessboard. $C = \{\text{red}, \text{white}, \text{blue}\}$ and we will take as our weight function ω defined by

$$\omega(\text{red}) = r, \qquad \omega(\text{white}) = w \qquad \text{and} \qquad \omega(\text{blue}) = b.$$

The group G is the group of permutations of the nine squaresqs which correspond to the eight symmetries of the square. It is easy to work out the cycle types of these permutations, and I give them in the table below. I have used our standard notation for the symmetries of the square.

e	a	b	c	p	q	r	s
x_1^9	$x_1x_4^2$	$x_1x_2^4$	$x_1x_4^2$	$x_1^3x_2^3$	$x_1^3x_2^3$	$x_1^3x_2^3$	$x_1^3x_2^3$

It follows that the cycle index of G is given by

$$C_G(x_1, x_2, x_4) = \tfrac{1}{8}(x_1^9 + x_1x_2^4 + 4x_1^3x_2^3 + 2x_1x_4^2).$$

So, by Pólya's Theorem, the pattern inventory is

$$\begin{aligned}\tfrac{1}{8}((r + w + b)^9 &+ (r + w + b)(r^2 + w^2 + w^2)^4 \\ &+ 4\,(r + w + b)^3\,(r^2 + w^2 + b^2)^3 \\ &+ 2(r + w + b)(r^4 + w^4 + b^4)^2).\end{aligned}$$

It would be rather tedious to expand this expression to obtain the coefficients of each term in the pattern inventory. (There are now computer programs that will do this sort of algebra for you.) To illustrate how it could be done we will evaluate just one of the terms.

We will work out how many different patterns there are with two red squares, three white squares and four blue squares. Thus we are seeking the coefficient of $r^2w^3b^4$ in the pattern inventory. We consider the four terms in the pattern inventory one by one.

By the Multinomial Theorem, the coefficient of $r^2w^3b^4$ in the term $(r + w + b)^9$ is $9!/2!3!4! = 1260$.

The only term involving $r^2w^3b^4$ in the term $(r + w + b)(r^2 + w^2 + w^2)^4$ comes by multiplying w from the first bracket by $r^2w^2b^4$ from the second bracket, whose coefficient, again by the Multinomial Theorem, is 12.

Dealing with the third term, $4(r + w + b)^3(r^2 + w^2 + b^2)^3$, is a little more complicated as it contributes terms involving $r^2w^3b^4$ in three different ways. We can obtain $r^2w^3b^4$ from the product of the two brackets in the following ways: multiplying w^3 from the first bracket by $3r^2b^4$ from the second, multiplying $3r^2w$ from the first bracket by $3w^2b^4$ from the second, and multiplying $3wb^2$ from the first bracket by $6r^2w^2b^2$ from the second. (Again we have used the Multinomial Theorem to obtain the coefficients of these terms.) Hence, not forgetting the four in front, the coefficient of $r^2w^3b^4$ in the term $4(r + w + b)^3(r^2 + w + b^2)^3$ is

$$4(1 \times 3 + 3 \times 3 + 3 \times 6) = 120.$$

After all this hard work it is a relief to note that the final term in the pattern inventory, namely $2(r + w + b)(r^4 + w^4 + b^4)^2$, does not include any terms involving $r^2w^3b^4$.

Thus the coefficient of $r^2w^3b^4$ is

$$\tfrac{1}{8}(1260 + 12 + 120) = 174,$$

and thus there are 174 different patterns involving two red, three white and four blue squares. This number is already sufficiently large that to solve this problem by a systematic listing of all the separate possibilities would not be very feasible.

Further examples of problems of this kind can be found in the Exercises below. Do try them for yourself before looking at the solutions.

Exercises

11.4.1. How many different patterns can be formed by colouring the squares of a 5×5 chessboard black and white, so that there are 15 black squares and ten white squares?

11.4.2. How many different patterns can be formed by colouring the triangles into which the square is divided so that there are two red triangles, two white triangles and four blue triangles?

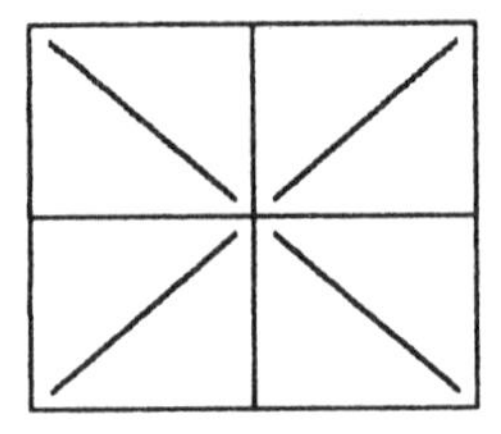

11.4.3. How many different patterns can be formed by colouring the faces of a cube so that there are three red faces, two white faces and one blue face?

11.4.4. How many different patterns can be formed by colouring the faces of a regular octahedron so that there are four red faces, two white faces and two blue faces?

11.4.5. How many different positions can arise in a game of noughts and crosses?

11.5 PÓLYA'S THEOREM: THE PROOF

We now set about giving a proof of Pólya's Theorem. Throughout this section we suppose that D and C are two sets, that ω is a weight function on C, and that G is a group of permutations of D which acts on the set X of all mappings from D to C, by the action given, by

$$\text{for } \pi \in G \quad \text{and} \quad f \in X,$$
$$\pi \cdot f = f \circ \pi^{-1}.$$

Our strategy is to partition X according to the weights of the mappings, to observe that G acts on each set of this partition separately, and to use Burnside's Theorem to count the number of different mappings of any given weight. Our proof proceeds by a series of lemmas.

When in section 10.2 we tackled the problem of the colourings of the 8×8 chessboard, we made use of the fact that a colouring was fixed by a particular symmetry if and only if all the squares in each cycle of the corresponding permutation had the same colour. The same principle was used in the other problems in that section.

In the general situation of this section, the equivalent property is that a colouring f is fixed by a permutation π if and only if f is constant on each cycle of π. That is for every cycle

$$(d_1\ d_2\ \ldots\ d_k)$$

of π, we must have

$$f(d_1) = f(d_2) = \ldots = f(d_k).$$

Note also that $(d_1\ d_2 \ldots d_k)$ is a cycle of π if and only if

$$d_2 = \pi(d_1),\ d_3 = \pi(d_2),\ \ldots \text{ and } d_1 = \pi(d_k).$$

Thus, writing d for d_1, it follows that the cycles of π have the form

$$(d\ \pi(d)\ \pi(\pi(d))\ \ldots).$$

It follows that f is constant on each cycle of π if and only if for all $d \in D$,

$$f(d) = f(\pi(d)) \tag{11.5}$$

We now give the general proof.

Lemma 11.3
Let π be a permutation of D. Then, for each $f \in X$,

$$f \in \text{Fix}(\pi) \Leftrightarrow f \text{ is constant on each cycle of } \pi.$$

Proof
We have, by the definition of the action of G on X, that

$$\begin{aligned} f \in \text{Fix}(\pi) &\Leftrightarrow \pi \cdot f = f \\ &\Leftrightarrow f \circ \pi^{-1} = f \\ &\Leftrightarrow \text{for all } d \in D,\ f(\pi^{-1}(d)) = f(d). \end{aligned} \tag{11.6}$$

Now as π is a permutation of D, $\{\pi^{-1}(d) : d \in D\} = D$, and hence it follows that, writing d' for $\pi^{-1}(d)$,

$$f \in \text{Fix}(\pi) \Leftrightarrow \text{for all } d' \in D,\ f(d') = f(\pi(d'))$$

and hence, by (11.5), that

$$f \in \text{Fix}(\pi) \Leftrightarrow f \text{ is constant on each cycle of } \pi.$$

The next lemma provides a formula for the inventory of Fix(π). By Lemma 11.3 this is the inventory of all the functions in X which are constant on each cycle of π. The sets of elements of D making up each cycle of π partition D into disjoint sets. So the next Lemma concerns itself with the inventory of all those functions which are constant on each set of such a partition.

Lemma 11.4
Let $D = D_1 \cup D_2 \cup \ldots \cup D_k$ be a partition of D into disjoint sets. Then the inventory of all those functions in X which are constant on each set D_i is

$$\sum_{c \in C} \omega(c)^{\#(D_1)} \times \sum_{c \in C} \omega(c)^{\#(D_2)} \times \ldots \times \sum_{c \in C} \omega(c)^{\#(D_k)}$$

Proof
Choosing a function, f, which is constant on each set D_i is equivalent to choosing, for $1 \leqslant i \leqslant k$, an element $c_i \in C$ to be the value of $f(d)$ for all $d \in D_i$. Since f takes this value $\#(D_i)$ times on D_i, such a choice of c_i contributes $\omega(c_i)^{\#(D_i)}$ to the weight of f. So the weight of f is obtained by choosing, for $1 \leqslant i \leqslant k$, one of the terms from the sum $\Sigma_{c \in C}$ $\omega(c)^{\#(D_i)}$ and then multiplying all the terms chosen in this way. Hence the sum of the weights of all the functions, f, which are constant on each of the sets D_i, consists of all the terms obtained in this way. That

is, the inventory of the set of all such functions is the product given above.

The formula given in Lemma 11.4 is beginning to look like the formula in Pólya's Theorem. The next lemma brings in the cycle types of the permutations. Recall that we are using $\mathrm{ct}(\pi)$ to stand for the cycle type of the permutation π.

Lemma 11.5

If $\mathrm{ct}(\pi) = x_{l_1}^{k_1} x_{l_2}^{k_2} \ldots x_{l_s}^{k_s}$, then the inventory of $\mathrm{Fix}(\pi)$ is

$$\left(\sum_{c \in C} \omega(c)^{l_1}\right)^{k_1} \times \left(\sum_{c \in C} \omega(c)^{l_2}\right)^{k_2} \times \ldots \times \left(\sum_{c \in C} \omega(c)^{l_s}\right)^{k_s}$$

Proof

By Lemma 11.3 a function f is in $\mathrm{Fix}(\pi)$ if and only if f is constant on each cycle of π. Suppose that $\mathrm{ct}(\pi) = x_{l_1}^{k_1} x_{l_2}^{k_2} \ldots x_{l_s}^{k_s}$. Then the cycles of π partition D into $k_1 + k_2 + \ldots + k_s$ sets, of which k_1 are of size l_1, k_2 of equal size l_2, and so on. Hence, by Lemma 11.4, the inventory of $\mathrm{Fix}(\pi)$ is as given.

Notice that the expression which occurs in Lemma 11.5 is obtained from the cycle type of π be replacing each occurrence of x_i by $\Sigma_{c \in C}\, \omega(c)^i$. We can indicate this by rewriting the expression as

$$\mathrm{ct}(\pi)\left(\sum_{c \in C} \omega(c),\ \sum_{c \in C} \omega(c)^2,\ \sum_{c \in C} \omega(c)^3,\ \ldots\right)$$

We now partition X into sets of functions of the same weight. The relation $\sim$ defined on X by

$$f \sim g \Leftrightarrow W(f) = W(g)$$

is an equivalence relation. We let $X_1, \ldots, X_t$ be the equivalence classes of this equivalence relation. Thus

$$X = X_1 \cup X_2 \cup \ldots \cup X_t, \text{ with } X_i \cap X_j = \varnothing \text{ for } i \neq j \qquad (11.7)$$

The next lemma shows that the group G acts on each set of this partition separately.

Lemma 11.6

Suppose that $f, g \in X$ have the same pattern (that is, f and g are in the same orbit of the group action). Then $W(f) = W(g)$.

Proof

Suppose f and g are in the same orbit. Then for some $\pi \in G$, $\pi \cdot f = g$. Thus $g = f \circ \pi^{-1}$ and hence,

$$W(g) = \prod_{d \in D} \omega(g(d)) = \prod_{d \in D} \omega(f(\pi^{-1}(d))). \qquad (11.8)$$

Now, as π is a permutation of D, the set of values taken by $\pi^{-1}(d)$, as d runs through D, is just D itself. That is

$$\{\pi^{-1}(d) : d \in D\} = \{d : d \in D\},$$

and hence

$$\prod_{d \in D} \omega(f(\pi^{-1}(d))) = \prod_{d \in D} \omega(f(d)) = W(f),$$

and so, by equation 11.8, $W(g) = W(f)$.

This lemma tells us that the group G acts on each set, X_i, of the partition (11.7) separately. That is, for each $\pi \in G$ and each $f \in X_i$, $\pi \cdot f \in X_i$. We can thus apply Burnside's Theorem to calculate the number of different orbits in X_i. This number gives us the coefficient of the corresponding term in the pattern inventory. We will use the notation π^i to indicate that we are restricting our attention to the action of π on X_i.

We are now ready to give the proof of Pólya's Theorem. For convenience we restate this theorem.

Theorem 11.2 *(Pólya's Theorem)*
Let X be the set of all mappings from the set D to the set C, and let ω be a weight function on C. Let G be a group of permutations of D which acts on X in the standard way. If the cycle index of G is $C_G(x_1, x_2, x_3, \ldots)$, then the pattern inventory is

$$C_G\left(\sum_{c \in C} \omega(c), \sum_{c \in C} \omega(c)^2, \sum_{c \in C} \omega(c)^3, \ldots\right).$$

Proof
Suppose that all the functions in the set X_i of this partition have weight W_i, and let there be m_i different patterns represented by the functions in X_i. By Lemma 11.6, no pattern is represented by functions in more than one set of this partition. Thus the pattern inventory, say PI, is given by

$$PI = \sum_{i=1}^{t} m_i W_i.$$

We have seen, from Lemma 11.5, that each $\pi \in G$ acts separately on each set X_i. It thus follows from Burnside's Theorem that

$$m_i = \frac{1}{\#(G)} \sum_{\pi \in G} \#(\mathrm{Fix}\,(\pi^i)$$

and hence

$$PI = \frac{1}{\#(G)} \sum_{i=1}^{t} \left(\sum_{\pi \in G} \#(\mathrm{Fix}\,(\pi^i)\right) W_i.$$

Both the sums in this last formula are finite sums, and hence we can interchange the order of summation to get

$$PI = \frac{1}{\#(G)} \sum_{\pi \in G} \left(\sum_{i=1}^{t} \#(\text{Fix}\,(\pi^i)) W_i \right). \tag{11.9}$$

Now $\Sigma_{i=1}^{t}\ \#(\text{Fix}\,(\pi^i))W_i$ is the inventory of $\text{Fix}\,(\pi)$, and hence by Lemma 11.5, it is equal to

$$\text{ct}\,(\pi) \left(\sum_{c \in C} \omega(c), \sum_{c \in C} \omega(c)^2, \sum_{c \in C} \omega(c)^3, \ldots \right)$$

and so

$$PI = \frac{1}{\#(G)} \sum_{\pi \in G} \text{ct}\,(\pi) \left(\sum_{c \in C} \omega(c), \sum_{c \in C} \omega(c)^2, \sum_{c \in C} \omega(c)^3, \ldots \right) \tag{11.10}$$

Adding up the expressions $\text{ct}\,(\pi)$ and then substituting $\Sigma_{c \in C}\ \omega(c)^i$ for x_i leads to the same expression as first making the substitutions and then adding. Hence, from equation (11.10),

$$PI = \left(\frac{1}{\#(G)} \sum_{\pi \in G} \text{ct}\,(\pi) \right) \left(\sum_{c \in C} \omega(c), \sum_{c \in C} \omega(c)^2, \sum_{c \in C} \omega(c)^3, \ldots \right) \tag{11.11}$$

Now $(1/\#(G))\Sigma_{\pi \in G}\,\text{ct}\,(\pi)$ is the cycle index, C_G, of G. Thus we can deduce from equation (11.11) that

$$PI = C_G \left(\sum_{c \in C} \omega(c), \sum_{c \in C} \omega(c)^2, \sum_{c \in C} \omega(c)^3, \ldots \right).$$

11.6 COUNTING SIMPLE GRAPHS

In this section we show how Pólya's Theorem can be used to count the number of different graphs with a specified number of vertices. The computation, though in principle straightforward, is somewhat long. We illustrate the method for the easy case of graphs with three vertices, and for graphs with five vertices, to show how long this takes. Thus we leave to the reader the case of graphs with four vertices, where calculation by hand is feasible, and, for brave readers, the case of graphs with six vertices.

Recall, first, that we are using *graph* to mean a graph without multiple edges, and without loops, that is, what is often referred to as a *simple graph*. If multiple edges are allowed, then even with just two vertices there are infinitely many different graphs, as the two vertices could be joined by any finite number of edges.

In Chapter 9 we described graphs as consisting of a pair (V, E), where V is the set of vertices, and E is an irreflexive and symmetric relation defined on V. From this point of view, two vertices u and v are joined by an edge if and only if uEv. To bring Pólya's Theorem to bear we need to be able to view graphs in terms of colourings, that is, in terms of mappings from a set D to a set C. So we need to look at a graph from a slightly different point of view.

Consider, for example, the following graph with 5 vertices:

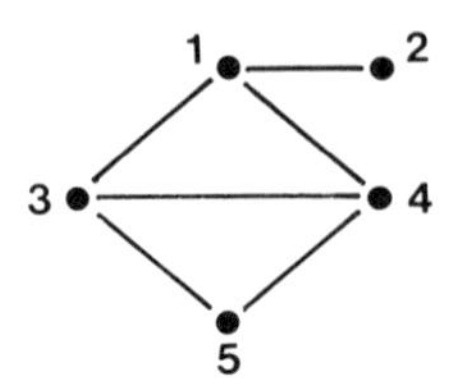

The vertex set of a graph can be any set we like. Here, when we are considering a graph with five vertices, it is convenient to suppose that the vertex set is the set of the first five positive integers $\{1, 2, 3, 4, 5\}$. The graph is determined by knowing for each two-element subset $\{i, j\}$ of $\{1, 2, 3, 4, 5\}$ whether the vertices i and j are joined in the graph or not. Hence the graph is described by a function, whose domain is the set of all these two-element sets. We let D_5 be this set, namely

$$\{\{1,2\}, \{1,3\}, \{1,4\}, \{1,5\}, \{2,3\}, \{2,4\}, \{2,5\}, \{3,4\}, \{3,5\}, \{4,5\}\}$$

A graph with 5 vertices can thus be described by a mapping

$$f: D_5 \to \{0, 1\}$$

such that

$$f(\{i, j\}) = 1 \Leftrightarrow i \text{ and } j \text{ are connected by an edge.} \qquad (11.12)$$

Thus the graph with five vertices pictured above corresponds to the mapping f as defined by the following table:

$\{i, j\}$	$f(\{i, j)\})$
$\{1, 2\}$	1
$\{1, 3\}$	1
$\{1, 4\}$	1
$\{1, 5\}$	0
$\{2, 3\}$	0
$\{2, 4\}$	0
$\{2, 5\}$	0
$\{3, 4\}$	1
$\{3, 5\}$	1
$\{4, 5\}$	1

In general, for each positive integer, n, we let D_n be the set of all two-element subsets of the set $\{1, 2, \ldots, n\}$. We will call D_n the set of **possible edges**. A graph with n vertices is defined by a mapping

$$f: D_n \to C, \text{ where } C = \{0, 1\},$$

which specifies via condition (11.12) which pairs of vertices are connected by an edge.

The most convenient weight function to use is ω, defined by

$$\omega(0) = 1$$

$$\omega(1) = c$$

as it will then follow that a graph with k edges has weight c^k. (Although we have chosen only to consider simple graphs, it is worth noting that this way of viewing graphs as colourings of the set of possible edges can easily be extended to deal with graphs with multiple edges. All we need do is to replace the set C by the set of natural numbers and replace conditions (11.12) by

$$f(\{i, j\}) = k \Leftrightarrow i \text{ and } j \text{ are connected by } k \text{ edges.})$$

Next we have to decide which is the relevant group of permutations of the set D_n of possible edges. Consider a straightforward example of a pair of isomorphic graphs:

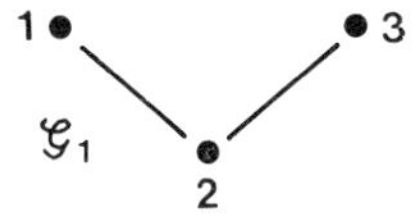

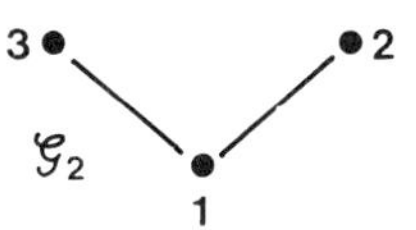

These graphs correspond to the colourings f_1, f_2 shown below.

$\{i, j\}$	$f_1(\{i, j\})$
$\{1, 2\}$	1
$\{1, 3\}$	0
$\{2, 3\}$	1

$\{i, j\}$	$f_2(\{i, j\})$
$\{1, 2\}$	1
$\{1, 3\}$	1
$\{2, 3\}$	0

The mapping $\pi: \{1, 2, 3\} \to \{1, 2, 3\}$ defined by

$$\pi(1) = 3$$

$$\pi(2) = 1$$

$$\pi(3) = 2$$

is an isomorphism between these two graphs. We need to reinterpret this isomorphism in terms of our new way of viewing graphs as colourings of the set D_n of possible edges. Corresponding to the permutation π of the vertices there is a permutation π^* of the possible edges. Since $\pi(1) = 3$ and $\pi(2) = 1$, the possible edge $\{1, 2\}$ is mapped to the possible edge $\{3, 1\}$, which, of course, is the same as $\{1, 3\}$. The following table shows the permutation of the possible edges corresponding to the permutation π:

$$\{1, 2\} \to \{1, 3\}$$

$$\{1, 3\} \to \{2, 3\}$$

$$\{2, 3\} \to \{1, 2\}$$

We use π^* to stand for this permutation of the possible edges. Comparing this table with those of the colourings f_1 and f_2 above, we see that, because π is an isomorphism between the two graphs, possible edges that are matched by π^* are assigned the same colours by f_1 and f_2. In symbols

$$f_2(\pi^*(\{i, j\})) = f_1(\{i, j\}), \text{ for all } \{i, j\} \in D_3$$

and thus

$$f_1 = f_2 \circ \pi^*.$$

It follows that

$$f_2 = f_1 \circ \pi^{*-1}. \tag{11.13}$$

Equation (11.13) should be compared with equation (11.1). We see from it that the action of π^* on the colourings which define graphs is exactly the same as the group action involved when we were considering colourings of chessboards and other figures.

We are now ready to generalize this discussion to deal with the general case. Given a permutation π of $\{1, 2, 3, \ldots, n\}$, that is, $\pi \in S_n$, there corresponds a permutation, which we write as π^*, of D_n. π^* is defined by

$$\pi^*(\{i, j\}) = \{\pi(i), \pi(j)\}, \text{ for } \{i, j\} \in D_n.$$

Thus the group we are working with is the group of all the permutations π^* corresponding to permutations $\pi \in S_n$. We let S_n^* be this group. Thus

$$S_n^* = \{\pi^* : \pi \in S_n\}.$$

The group S_n^* acts on the set X of mappings from D_n to $C = \{0, 1\}$ in the standard way. That is, for $\pi^* \in S_n^*$ and $f \in X$,

$$\pi^* \cdot f = f \circ \pi^{*-1}.$$

In Exercise 11.6.6 you are asked to prove that, for $n \neq 2$, the mapping $\pi \mapsto \pi^*$ is an isomorphism from S_n to S_n^*. It follows that, for $n \neq 2$, S_n^* contains $n!$ permutations. The set D_n contains $n(n-1)/2$ possible edges and thus S_n^* consists, in general, of only a very small proportion of all the permutations of D_n.

We can now calculate the pattern inventory of the graphs with n vertices by working out the cycle index of the group S_n^* and making the appropriate substitutions. As promised above, we give the calculation for the cases $n = 3$ and $n = 5$.

Example 3: Graphs with three vertices

In calculations of the kind we are about to do, it is convenient to simplify our notation for sets, so that we do not get cluttered up with the symbols { and }. So instead of writing

$$D_3 = \{\{1, 2\}, \{1, 3\}, \{2, 3\}\},$$

we will simply write

$$D_3 = \{12, 13, 23\}.$$

This notation is very convenient provided that $n \leqslant 9$, but we will need to remember that, for example, in this context '12' does not mean 'twelve', but instead denotes the two-element set $\{1, 2\}$. In this notation '21' denotes the same two-element set.

We need first to calculate the cycle index of the group S_3^*. To do this we have to take the permutations in S_3 and work out the cycle types of the corresponding permutations in S_3^*. We do not have to do this for each element of S_3 separately. If two permutations, π_1 and π_2, have the same cycle type, then so too do the associated permutations π_1^* and π_2^* (you are asked to prove this, in general, in Exercise 11.6.4). So we need only consider one permutation of each cycle type from S_3, which we do as follows.

If π is the identity element of S_3, of cycle type x_1^3 then, clearly, π^* is the identity element of S_3^*. Exceptionally, $\#(D_3) = 3$, and so in this case π^* has the same cycle type as π.

Next, suppose that π has cycle type x_1x_2. Say $\pi = (1\,2)$. Then in bracket notation π^* is

$$\begin{pmatrix} 12 & 13 & 23 \\ 21 & 23 & 13 \end{pmatrix}.$$

We obtained this bracket notation for π^* by listing the elements of D_3 in the top line, and then writing under each of the numbers 1, 2, 3 their images under the permutation π. Since 21 is the same set as 12 (namely, $\{1, 2\}$), we can write π^* in cycle notation as

$$\pi^* = (13\ 23)$$

and hence π^* also has cycle type x_1x_2. Do not jump too quickly to the wrong conclusion. The case $n = 3$ is special, as you will see from the calculation with $n = 5$ in the next example.

Finally, we consider the case where π has cycle type x_3. Say $\pi = (1\,2\,3)$. Then

$$\pi^* = \begin{pmatrix} 12 & 13 & 23 \\ 23 & 21 & 31 \end{pmatrix} = (12\ 23\ 13)$$

and hence also has cycle type x_3 (another coincidence!).

We can summarize these calculations in the following table:

Cycle type of $\pi \in S_3$	*No. of permutations of this cycle type*	*Cycle type of* $\pi^* \in S_3^*$
x_1^3	1	x_1^3
x_1x_2	3	x_1x_2
x_3	2	x_3

Thus it follows that the cycle index of S_3^* is

$$\tfrac{1}{6}(x_1^3 + 3x_1x_2 + 2x_3).$$

Again we must emphasize that it is only a coincidence that this is the same as the cycle index of S_3. Using Pólya's Theorem, we can now obtain the pattern inventory by replacing x_1 by $1 + c$, x_2 by $1^2 + c^2$ (i.e. $1 + c^2$) and x_3 by $1 + c^3$, Thus it is

$$\tfrac{1}{6}((1 + c)^3 + 3(1 + c)(1 + c^2) + 2(1 + c^3)).$$

If all we want to know is the total number of different graphs then, by putting $c = 1$, we get the sum of the coefficients in the pattern inventory. If we want to classify these graphs according to the number of edges, then we need to expand the pattern inventory to obtain the coefficient of each term. Doing this, we obtain

$$1 + c + c^2 + c^3,$$

showing that there is one graph with no vertices, and one with each of 1, 2 and 3 vertices. Hence altogether there are 4 different graphs with 3 vertices. It is straightforward to list them.

The arithmetic in this example was not too heavy, but when we increase the number of vertices the work becomes a lot harder.

Example 4: Graphs with five vertices

We assume that the vertex set of a graph with five vertices is $\{1, 2, 3, 4, 5\}$. Then the set D_5 of possible edges has ten elements. Using our abbreviated notation, we can write D_5 as

$$D_5 = \{12\ 13\ 14\ 15\ 23\ 24\ 25\ 34\ 35\ 45\}.$$

As in the previous example, we need to take permutations, π, of all the different cycle types in S_5, and for each of them work out the cycle type of the corresponding permutation π^*. We illustrate just one example in detail of these calculations, with a permutation, π of cycle type x_2x_3.

We take $\pi = (1\,2\,3)(4\,5)$. As before our method is to first work out π^* in bracket notation.

$$\pi^* = \begin{pmatrix} 12 & 13 & 14 & 15 & 23 & 24 & 25 & 34 & 35 & 45 \\ 23 & 21 & 25 & 24 & 31 & 35 & 34 & 15 & 14 & 54 \end{pmatrix}$$

$$= (12\ 23\ 13)(14\ 25\ 34\ 15\ 24\ 35)(45).$$

(The most efficient way to do this calculation is first to list the elements of D_5 in the top row. Since $\pi(1) = 2$ we then, in the second row write a 2 under each 1, since $\pi(2) = 3$, we then write 3 under each 2 and so on. The final step is to rewrite π^* in cycle form.)

Thus π^* has cycle type $x_1x_2x_3$. Repeating this calculation for the other cycle types we obtain the following table (which you are encouraged to check for yourself):

Cycle type of $\pi \in S_5$	*No. of permuations of this cycle type*	*Cycle type of* $\pi^* \in S_5^*$
x_1^5	1	x_1^{10}
$x_1^3x_2$	10	$x_1^4x_2^3$
$x_1^2x_3$	20	$x_1x_3^3$
$x_1x_2^2$	15	$x_1^2x_2^4$
x_1x_4	30	$x_2x_4^2$
x_2x_3	20	$x_1x_3x_6$
x_5	24	x_5^2

It follows that the cycle index of S_5^* is given by

$$C_{S_5^*} = \tfrac{1}{120}(x_1^{10} + 10x_1^4x_2^3 + 20x_1x_3^3 + 15x_1^2x_2^4 + 30x_2x_4^2 + 20x_1x_3x_6 + 24x_5^2)$$

Hence the pattern inventory for simple graphs with five vertices is

$$\begin{aligned}&\tfrac{1}{120}((1+c)^{10} + 10(1+c)^4(1+c^2)^3 + 20(1+c)(1+c^3)^3\\ &\quad + 15(1+c)^2(1+c^2)^4 + 30(1+c^2)(1+c^4)^2\\ &\quad + 20(1+c)(1+c^3)(1+c^6) + 24(1+c^5)^2)\\ &= 1 + c + 2c^2 + 4c^3 + 6c^4 + 6c^5 + 6c^6 + 4c^7 + 2c^8 + c^9 + c^{10}.\end{aligned}$$

From this we see that the number of different graphs with five vertices is as in the following table:

Edges	0	1	2	3	4	5	6	7	8	9	10
No. of graphs	1	1	2	4	6	6	6	4	2	1	1

Thus there are altogether 34 different graphs with five vertices. You are asked in Exercise 11.6.3 to list all these 34 graphs. It is hoped that this Exercise will convince you that Pólya's Theorem, despite all the calculations involved, provides a better method for calculating the number of different graphs than simply trying to list them.

It can be seen that, in principle, this calculation can be repeated for graphs with any finite number of vertices. The number of cycle types increases very rapidly as the number of vertices increases, so it becomes less and less feasible to do the calculations by hand. You might like to think about how to write a computer program to do all the hard work for you.

Exercises

11.6.1. Calculate the pattern inventory for graphs with four vertices.

11.6.2. Calculate the pattern inventory for graphs with six vertices.

11.6.3. List all the 34 different graphs with five vertices.

11.6.4. Prove that if π_1, π_2 are two permutations from S_n which have the same cycle type, then π_1^*, π_2^* also have the same cycle type.

11.6.5. Is the converse result true? That is, if π_1^*, π_2^* have the same cycle type, does it follow that π_1 and π_2 have the same cycle type?

11.6.6. Prove that, for $n > 2$, the mapping $\pi \mapsto \pi^*$, from S_n to S_n^* is an isomorphism.

11.7 CONCLUSION

Unless you are one of those readers who always starts a book from the end, you will presumably only be reading this if you have already read most of the preceding pages. And this, I hope, will be because you have found the mathematics contained in them interesting. The best books always end too soon, leaving the reader wanting more. This book has not quite finished yet because there are the solutions to most of the Exercises still to come. If you still want more, there follows after these solutions some supplementary exercises and some suggestions for further reading which I hope will enable you to explore more of the exciting world of combinatorics.

Supplementary exercises

These supplementary exercises are provided for readers who would like to tackle problems for which no solutions are provided. Most of them routine. I have used an * to mark questions which I think are quite hard, and ** for those which are very hard or for which I do not have a solution. The problems are arranged by chapters and the order within chapters corresponds roughly to the order of the material in each chapter.

CHAPTER 1

1. British £10 notes currently carry a serial number of the form

   CZ23 815492

   that is, two letters, followed by a two-digit number, followed by a six-digit number. How many different serial numbers of this form are there?
2. A drawer contains 20 pairs of socks, all of different designs. If socks are taken from the drawer, one by one, at random, how many socks have to be removed before there is a better than even chance that you have removed at least one matching pair of socks?
3. How many sequences are there of n digits whose sum is a multiple of 5?
4. A hand of seven cards is dealt at random from a standard pack of 52 cards. Calculate the ratio p/q, where p is the probability that the hand has a 3-3-1-0 suit distribution and q is the probability that it has a 4-2-1-0 suit distribution.
5. How many possible different suit distributions are there for a bridge hand of 13 cards?

6. Calculate how many bridge hands there are for each of the possible suit distributions.
7. It is customary to value a bridge hand by counting an ace as four points, a king as three points, a queen as two points and a jack as one (this is the *Milton Work point count*, named after the American bridge player Milton Work). The value of a bridge hand in this system can vary between 0 and 37 points. Calculate how many bridge hands there are with each of these point values.
8. If $6n$ dice are thrown simultaneously, what is the probability that each of the possible scores 1, 2, 3, 4, 5, 6, is obtained an equal number of times, i.e. each occurs on n of the dice?
9. Prove that for all positive integers k, m, n, with $m, n \geqslant 1$, $k \leqslant m$ and $k \leqslant n$,

$$\sum_{s=0}^{k} C(m, s)C(n, k-s) = C(m+n, k).$$

10. *Generalize the result of Exercise 1.4.7 to deal with the case where there are balls of k colours in the bag and n balls are sampled with replacement, $n \leqslant k$. Show that the probability that at least two of the sampled balls have the same colour is a minimum when there are equal numbers of balls of each colour. This justifies the remark in Exercise 1.4.6 that if birthdays are not equally distributed throughout the year, then the probability of a coincidence increases.

CHAPTER 2

1. How many integers are there in the range from 1 to 10^{30} which are perfect squares, or perfect cubes or perfect fifth powers?
2. How many integers are there in the range from 1 to 10^6 which are divisible by 2 or 3 or 5 or 7?
3. A bag contains equal numbers or red, green, blue, and yellow balls. One ball is drawn at random from the bag and then replaced. If this is done eight times, what is the probability that at least one ball of each colour is drawn from the bag?
4. **Find a general formula for the number of *anagrams* (as defined in Exercise 2.2.3) of a word in terms of the number of repetitions of letters within the word.

CHAPTER 3

1. Let $u_k(n)$ be the number of partitions of n into unequal parts (i.e. all the numbers in the partition must be different) of size at most k. Prove that for all $k, n \in \mathbb{N}$, with $k \leqslant n$,

$$u_k(n) = u_{k-1}(n) + u_{k-1}(n-k).$$

2. Prove that the number of partitions of n into unequal odd parts is the same as the number of partitions of n which are self-dual (i.e. identical with their dual).
3. **Prove that if $n \equiv 4 \pmod 5$ then $p(n) \equiv 0 \pmod 5$.
4. Prove that for all $n \in \mathbb{N}^+$,

$$p(n^2) \geqslant \frac{n^{2n-1}}{(n!)^2}.$$

CHAPTER 4

1. Determine which pairs of the following functions are asymptotic.
   (a) $f_1 : n \mapsto n^2$, (b) $f_2 : n \mapsto \sin(n^2)$,
   (c) $f_3 : n \mapsto n^2 + \sin(n^2)$ (d) $f_4 : n \mapsto 2^n$,
   (e) $f_5 : n \mapsto n^2 + 2^n$, (f) $f_6 : n \mapsto 2^{2^n}$,
   (g) $f_7 : n \mapsto \dfrac{n^5 + 3n^2}{n^3 + 6n}$, (h) $f_8 : n \mapsto \ln_2(1 + 2^{2^n})$.
2. Estimate the value of

$$\frac{(2\pi)^{100}1000!}{(10!)^{100}\sqrt{\pi}}$$

   correct to 3 significant figures.
3. Use Stirling's Approximation for $n!$ to determine whether or not the binomial coefficient $C(200, 100)$ is larger than $C(100, 50)^2$.

CHAPTER 5

1. For each positive integer k, let $x \to F_k(x)$ be the generating function for the sequence $\{n^k\}$.
   (a) Prove that for each positive integer k

$$F_{k+1}(x) = xF'_k(x),$$

   where the $'$ denotes differentiation with respect to x.
   (b) Hence deduce that $F_k(x)$ can be expressed in the form

$$\frac{p_k(x)}{(1-x)^{k+1}}$$

   where $p_k(x)$ is a polynomial of degree k in which the coefficient of x^k is 1.
2. Let s be a positive integer. Let $a_s(n)$ be the number of partitions of n in which no part is repeated more than s times, and let $b_s(n)$ be the number of partitions of n in which no part is a number divisible by $s+1$. Find the generating functions for the sequences $\{a_s(n)\}$ and $\{b_s(n)\}$, and hence deduce that for each positive integer n,

$$a_s(n) = b_s(n)$$

Can you give a combinatorial proof of this result?

3. *Let P be the generating function for the sequence $p(n)$ of unrestricted partition numbers and for each positive integer k let P_k be the generating function for the sequence $\{p_k(n)\}$. Prove that

(a)
$$P(x) = \sum_{k=1}^{\infty} x^{k^2}(P_k(x))^2$$

and (b)
$$P(x) = \sum_{k=0}^{\infty} x^{k(k+1)} P_k(x) P_{k+1}(x)$$

CHAPTER 6

1. Find explicit formulae for each of the sequences $\{a_n\}$ as defined by the following recurrence relations.

(a)
$$a_{n+2} = 5a_{n+1} - 6a_n \quad \text{for } n \geqslant 1,$$

with $a_1 = 1$ and $a_2 = 2$.

(b)
$$a_{n+2} = 6a_{n+1} - 9a_n \quad \text{for } n \geqslant 1,$$

with $a_1 = 1$ and $a_2 = 2$.

(c)
$$a_{n+3} = 5a_{n+2} + 4a_{n+1} - 20a_n \quad \text{for } n \geqslant 1,$$

with $a_1 = 1$, $a_2 = 2$ and $a_3 = 3$.

2. The sequence $\{a_n\}$ is defined by

$$a_{n+2} = 2a_{n+1} + a_n + (-1)^n, \text{ for } n \geqslant 1, \text{ with } a_1 = 0 \text{ and } a_2 = 2.$$

Find the generating function for the sequence $\{a_n\}$ in as simple a form as possible. Hence find an explicit formula for a_n and a constant α such that a_n is asymptotic to $\alpha 2^n$.

3. The sequence $\{a_n\}$ is defined by

$$a_{n+2} = 2a_{n+1} + 8a_n - 9n, \text{ for } n \geqslant 1, \text{ with } a_1 = 1 \text{ and } a_2 = 3.$$

Find the generating function for the sequence $\{a_n\}$ in as simple a form as possible, and hence or otherwise, find an explicit formula for a_n.

4. Let a_n be the number of n-digit sequences formed using only the integers 0, 1, 2, 3, in which 0 occurs an odd number of times.

(a) Show that

$$a_{n+1} = 2a_n + 4^n$$

(b) Find the generating function for the sequence $\{a_n\}$ in the form

$$x \mapsto \frac{p(x)}{q(x)},$$

where $p(x)$, $q(x)$ are polynomials and hence, or otherwise, find an explicit formula for a_n.

5. Let a_n be the number of sequences of 0's, 1's and 2's in which a 0 can only be followed by a 1.
Find the recurrence relation satisfied by the sequence $\{a_n\}$ and hence find the generating function for the sequence $\{a_n\}$ in the form

$$x \mapsto \frac{p(x)}{q(x)},$$

where $p(x)$, $q(x)$ are polynomials. Hence or otherwise, find an explicit formula for a_n.

6. The sequence a_n is defined by the recurrence relation

$$a_n = 3 \sum_{s=0}^{n-1} a_s a_{n-1-s}, \text{ for } n \geqslant 1, \text{ with } a_0 = 2.$$

Find an explicit formula for a_n.

7. Express each of the following in partial fraction form.

(a) $\dfrac{1}{(x+2)(x-4)}$, (b) $\dfrac{x^2+1}{x^2-1}$,

(c) $\dfrac{1}{x^4-16}$, (d) $\dfrac{1}{(x^2-4)^2}$.

CHAPTER 7

1. Express the following permutation in disjoint cycle form.

$$\begin{pmatrix} 1 & 2 & 3 & 4 & 5 & 6 & 7 & 8 & 9 & 10 & 11 & 12 \\ 5 & 7 & 12 & 10 & 8 & 6 & 3 & 1 & 2 & 4 & 11 & 9 \end{pmatrix}.$$

2. Let $\sigma = (1\,2\,4)(3\,9)(5\,6\,7)$ and $\tau = (1\,8\,5\,3)(2\,6)(4\,9)$ be two permutations from S_9. Express the permutation $\tau\sigma\tau^{-1}$ in disjoint cycle form.
3. If a permutation is chosen at random from S_{2n} what is the probability that it is made up of two disjoint cycles each of length n?
4. Prove that if $(G, *)$ is a group with identity element e and $a, x, y \in G$ are such that $a*x = e = y*a$, then $x = y$.
5. Prove that if G is a non-empty set which is closed under the operation $*$ which satisfies:
(i) the Associativity Property: for all $x, y, z \in G$,

$$(x*y)*z = x*(y*z),$$

(ii) For all $a, b \in G$, there exists $x \in G$ such that

$$a * x = b,$$

and

(iii) For all $a, b \in G$, there exists $y \in G$ such that

$$y * a = b,$$

then $(G, *)$ is a group.

6. Draw up the Cayley table for the symmetry groups of
   (a) a rectangle (i.e. not a square),
   (b) a cuboid with rectangular faces.
7. Find plane figures whose symmetry groups are isomorphic to

$$\text{(a) } \mathbb{Z}_3, \quad \text{(b) } \mathbb{Z}_4, \quad \text{(c) } \mathbb{Z}.$$

8. *Find all the subgroups of S_5.
9. Prove that a group of order 6 must contain at least one element of order 2.
10. Prove that in finding the permutation in S_n of largest order it is sufficient to consider permutations in which the lengths of the cycles are either 1 or a prime number or a power of a prime number.
11. **Let $s(n)$ be the order of a permutation of greatest possible order in S_n. Prove that for each prime number, p, there is an integer N_p such that for all $n \geqslant N_p$, $s(n)$ is divisible by p.

CHAPTER 8

1. In each of the following cases determine whether the Group Action axioms hold.

(a) $$G = \left\{ \begin{pmatrix} a & b \\ 0 & a \end{pmatrix} : a, b \in \mathbb{R}, a \neq 0 \right\}$$

with the operation of matrix multiplication. $X = \mathbb{R}^2$.

$$\begin{pmatrix} a & b \\ 0 & a \end{pmatrix} \cdot (x, y) = (ax + by, ay)$$

(b) $$G = \left\{ \begin{pmatrix} a & b \\ 0 & c \end{pmatrix} : a, b, c \in \mathbb{R}, ac \neq 0 \right\}.$$

with the operation of matrix multiplication. $X = \mathbb{R}^2$.

$$\begin{pmatrix} a & b \\ 0 & c \end{pmatrix} \cdot (x, y) = (ax + b, cy)$$

(c) $G = \mathbb{Z}$ with the operation of addition. $X = \mathbb{Q}$. $n \cdot q = n + q$.

(d) $G = \mathbb{Q}$ with the operation of addition. $X = \mathbb{C}$. $q \cdot z = e^{q\pi i} z$.

2. In each of the above cases where the Group Action axioms hold determine the orbits of the action.
3. Suppose that the group G acts on the set X. The kernel of the action is the set

$$\{g \in G : \text{for all } x \in X,\ g \cdot x = x\}.$$

   (a) Prove that the kernel is a subgroup of G.
   (b) Determine the kernel in each of the cases in Exercise 1, above, where the Group Action axioms are satisfied.
   (c) Why is the kernel of a group action so called?

CHAPTER 9

1. **We saw in Chapter 9, Lemma 9.1 that the sum of the degrees of the vertices in a graph must be even. This is not the only restriction on the degrees of the vertices. You are asked to investigate this problem.
   More precisely, let k be a positive integer, and let $d_1, \ldots, d_k$ be a sequence of k positive integers. We say that this is a *degree sequence*, if there is a graph with k vertices, $v_1, \ldots, v_k$, such that for $1 \leqslant i \leqslant k$, $\delta(v_i) = d_i$. You are asked to investigate necessary and sufficient conditions for $d_1, \ldots, d_k$ to be a degree sequence.
2. List all the graphs with 5 vertices. For each of these graphs determine how many labellings it has.
3. A graph is said to be **planar** if it can be drawn in the plane so that the edges only cross at vertices. When a planar graph is drawn in the plane the regions into which it divides the plane are called **faces**. A well-known theorem due to Euler says that if a graph with v vertices and e edges is drawn in the plane with f faces (counting the unbounded region as one face) then

$$v + f = e + 2$$

   This formula is illustrated in the following example:

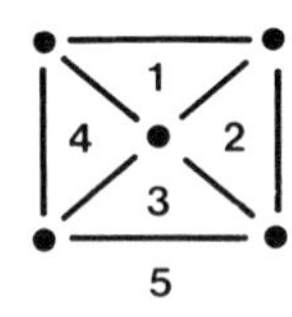

   In this example $v = 5$, $e = 8$ and as shown, $f = 5$. Thus $v + f = 10 = e + 2$.
   Prove that if $\mathcal{G}$ is a planar graph with v vertices and e edges, then

$$e \leqslant 3v - 6.$$

(Hint, count the edges in terms of the faces and use Euler's Theorem.)
Deduce that the graph, K_5, with 5 vertices, and in which each pair of vertices is joined by an edge is not planar.

4. *The graph $K_{3,3}$ with vertex set $V = \{a_1, a_2, a_3, b_1, b_2, b_3\}$, in which each vertex a_i is joined to each vertex b_j and there are no other edges is not planar although it satisfies the inequality $e \leqslant 3v - 6$ of Question 3, above. Modify the argument of that question to prove that $K_{3,3}$ is not planar.

CHAPTER 10

1. How many different patterns can be formed by assembling 27 black and white cubes to form a $3 \times 3 \times 3$ cube? (Consider only the rotational symmetries of the cube).

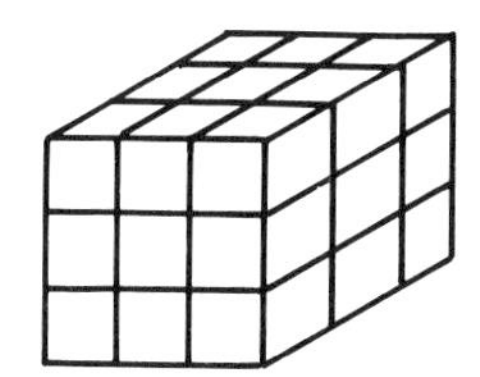

2. How many different patterns can be formed by colouring the small squares in the following figure using c colours? (Consider both the rotations and reflections of the figure).

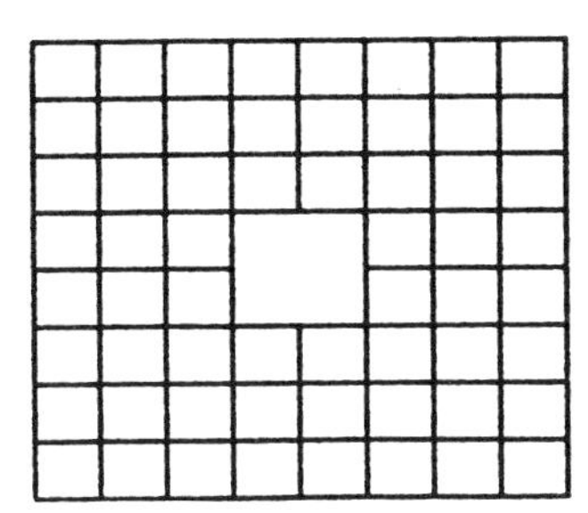

CHAPTER 11

1. Let D be a set of 10 elements, let C be a set of 3 elements, and let X be the set of all mappings from D to C. Let G be a group of permutations of D which acts on X by the rule:

$$\text{for } \pi \in G \text{ and } f \in X, \pi \cdot f = f \circ \pi^{-1}.$$

Prove that if $\#(G) = 49$ then at least 4 of the orbits contain just one element of X.

2. A group of twenty people are to travel from Dover to Calais. They have available eight train tickets (using the Channel tunnel, which we assume has been built), seven boat tickets and five aeroplane tickets. Each train ticket can be used on any of twelve trains, each boat ticket can be used on any of ten boats, and each aeroplane ticket can used on any of six planes. In how many different ways can the group make the journey?
3. How many different patterns can be formed by colouring the small squares in the following figure so that 30 are white and 30 are black? (Consider both rotations and reflections.)

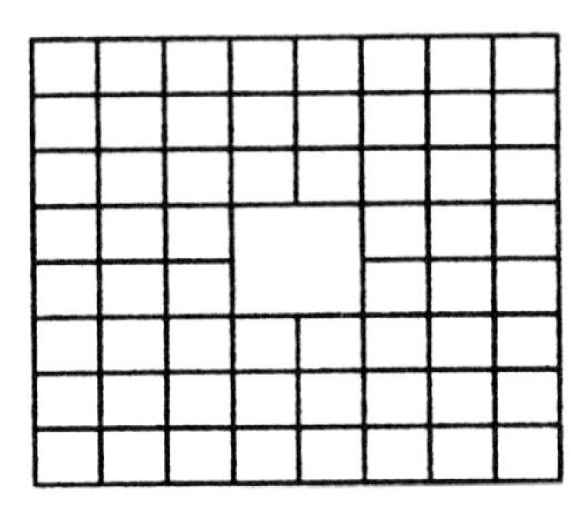

4. A $3 \times 3 \times 3$ cube is to be assembled from 27 small cubes of which one is black, eight are red, nine are white and nine are blue. The central cube must be black. How many different patterns can be formed. (Consider only the rotational symmetries of the cube.)
5. *Find a formula for the cycle type of $\pi^* \in S_n^*$ in terms of the cycle type of $\pi \in S_n$.

Solutions

CHAPTER 1

1.2.1. The initial letter can be chosen in 26 ways, the number in 999 ways (assuming that 0 is not possible) and the final sequence of 3 letters can be chosen in 26^3 ways. So the total number of difference licence plates is $26 \times 999 \times 26^3 = 456\,519\,024$.

1.2.2. In the first race with ten horses, the first three horses can be selected in $10 \times 9 \times 8 = 720$ ways; in the second race with eight horses they can be selected in $8 \times 7 \times 6 = 336$ ways; and in the six-horse race the first three horses can be selected in $6 \times 5 \times 4 = 120$ ways. Hence the total number of different ways in which these selections can be made is $720 \times 336 \times 120 = 29\,030\,400$ ways.

1.2.3. The first digit can be any of the ten possible digits, but since consecutive digits must be different, each subsequent digit can be chosen in just nine ways. Hence the total number of sequences of n digits in which no two consecutive digits are the same is

$$10 \times 9 \times 9 \times \ldots \times 9 = 10 \times 9^{n-1}.$$

1.3.1. The three pure mathematics options can be chosen from the 10 courses available in $C(12, 3) = 220$ ways. Likewise the two applied mathematics options can be chosen in $C(10, 2) = 45$ ways, the two statistics options in $C(6, 2) = 15$ ways, and the one computing option in $C(4, 1) = 4$ ways. So the total number of different ways in which the students can choose courses is $220 \times 45 \times 15 \times 4 = 594\,000$.

1.3.2. Each set of four points on the circumference of the circle gives rise to a pair of lines which meet inside the circle, and every point of intersection inside the circle arises in this way. Thus the maximum number of points of intersection that can be obtained is the number of ways of choosing four points from the n points on the circumference, that is $C(n, 4) = n(n-1)(n-2)(n-3)/24$. (It is always possible to arrange the n points so that this maximum is attained.)

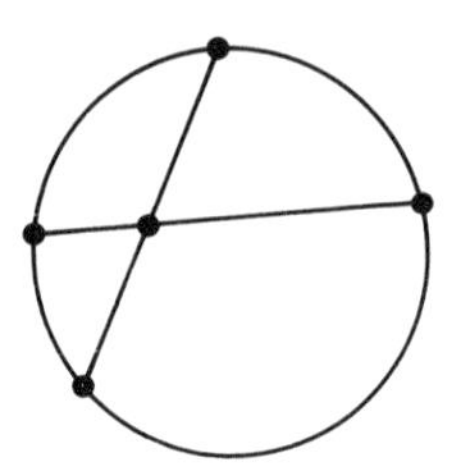

1.3.3. Suppose X is a set containing n elements. In how many ways can we choose a subset of X? For each element there are two choices, either to include it in the subset or not. So by the Principle of Multiplication of Choices (Theorem 1.1) there are altogether 2^n ways in which we can choose a subset of X. Hence this is the number of subsets X has. $C(n, r)$ gives the number of subsets of X containing r elements. Each subset of X contains r elements for some integer r between 0 and n. Hence altogether X has

$$\sum_{r=0}^{n} C(n, r)$$

subsets. So this sum must be equal to 2^n.

1.3.4. Suppose X is a set containing $2n$ elements. Split X up into disjoint sets, A and B, each containing n elements. We can select an n-element subset of X by choosing an integer r between 0 and n, and then choosing r elements from A and $n-r$ elements from B. Each sequence of these choices leads to a different n-element subset of X and each n-element subset of X can be obtained in this way. Hence altogether X has

$$\sum_{r=0}^{n} (C(n, r) \times C(n, n-r))$$

n-element subsets. Now, by Theorem 1.5, $C(n, n-r) = C(n, r)$. Also, we know that the total number of n-element subsets of a set with $2n$ elements is $C(2n, n)$. It follows that

$$\sum_{r=0}^{n} (C(n, r))^2 = C(2n, n).$$

1.3.5. Suppose that we have a set X of n objects. The product $C(n, k)C(k, s)$ counts the number of ways of first choosing a set, say Y, of k elements from X, and then choosing a subset of Y, say Z, containing s elements. We thus arrive at the following picture:

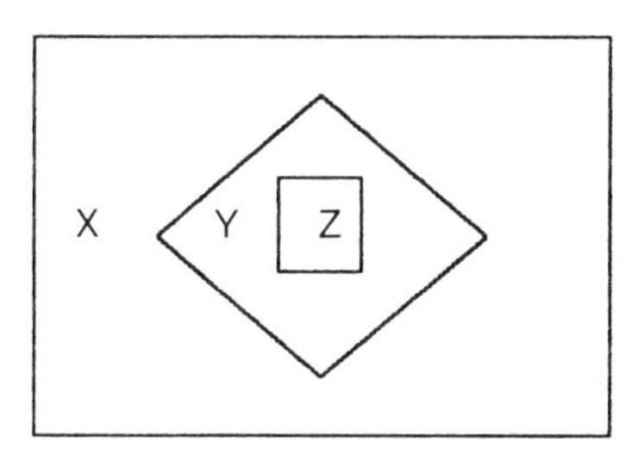

We can obtain exactly the same result by first choosing the subset Z consisting of s elements from X, and then choosing $k - s$ elements from the $n - s$ elements of $X \backslash Z$ to form $Y \backslash Z$. Z can be chosen from X in $C(n, s)$ ways and then $Y \backslash Z$ can be chosen from $X \backslash Z$ in $C(n - s, k - s)$ ways. So altogether this can be done in $C(n, s)C(n - s, k - s)$ ways.

Since these two expressions count the number of ways of arriving at the same situation they must be equal, that is

$$C(n, k)C(k, s) = C(n, s)C(n - s, k - s).$$

1.3.6. Suppose that we have a set

$$X = \{x_1, x_2, \ldots, x_{n-1}, x_n\}$$

containing n elements. We have seen in the solution to Exercise 1.3.3 that we can choose a subset of X in 2^n ways because for each of the elements $x_1, x_2, \ldots, x_{n-1}, x_n$ we have two choices, either to include it or to exclude it. If we wish to select a subset of X containing an even number of elements we still have, for $1 \leqslant i \leqslant n - 1$, a free choice as to whether or not to include x_i.

However, when we reach x_n we no longer have a choice. If the number of elements we have already decided to include in the subset is even, then we are forced to exclude x_n, to ensure that the resulting subset contains an even number of elements. If the number of elements we have already decided to include is odd, then we must include x_n so as make the total number of elements in the subset even. Hence, in selecting a subset from X which contains an even number of elements we can choose either to include or exclude an element on $n - 1$ occasions, but when it comes to the last element there is no choice. So the number of subsets of X containing an even number of elements is

$$2 \times 2 \times \ldots \times 2 \times 1 = 2^{n-1}$$

It follows that the number of subsets which contain an odd

number of elements is $2^n - 2^{n-1} = 2^{n-1}$. Thus the number of subsets containing an even number of elements is equal to the number of subsets containing an odd number of elements.

It follows that for each positive integer n,

$$\sum_{\substack{r=0 \\ r \text{ even}}}^{n} C(n, r) = \sum_{\substack{r=1 \\ r \text{ odd}}}^{n} C(n, r),$$

whence, as $(-1)^r$ is $+1$ for r even, and -1 for r odd,

$$\sum_{r=0}^{n} (-1)^r C(n, r) = 0.$$

1.3.7. The choice of a term of degree n involving k symbols corresponds to splitting a row of n objects into k blocks. For example, we consider the case $n = 9$ and $k = 4$. The term

$$x_1^2 x_3^4 x_4^3$$

of degree 9 in the symbols x_1, x_2, x_3 and x_4, corresponds to splitting up a row of nine objects as follows:

●●| |●●●●|●●●

The markers, |, divide up the row of nine bullets into blocks of lengths 2, 0, 4 and 3, corresponding to the fact that the degrees of x_1, x_2, x_3 and x_4 in the term are 2, 0, 4 and 3, respectively.

Clearly, in general, there is a one-to-one correspondence between terms of degree n involving k symbols, and rows of n objects divided into k blocks by $k - 1$ markers. Such a row thus consists of $n + k - 1$ symbols of which $k - 1$ are markers and n are bullets. So the number of such rows is the number of ways of choosing which $k - 1$ of the $n + k - 1$ symbols are to be markers. Thus there are $C(n + k - 1, k - 1)$ of them.

1.3.8. We need to count the number of ways we can choose x_1 from k_1 brackets, x_2 from k_2 brackets, and so on, when we expand the expression

$$(x_1 + x_2 + \ldots + x_s)(x_1 + x_2 + \ldots + x_s) \ldots (x_1 + x_2 + \ldots + x_s)$$

with n sets of brackets.

The k_1 brackets from which to choose x_1 can be selected in $C(n, k_1)$ ways. This leaves $n - k_1$ sets of brackets from which to choose x_2 and we can select k_2 of these brackets in $C(n - k_1, k_2)$ ways and so on. Thus the total number of ways of making these choices is $C(n, k_1)C(n - k_1, k_2) \ldots C(n - k_1 - k_2 - \ldots - k_{s-1}, k_s)$, which equals

$$\frac{n!}{k_1!(n-k_1)!} \times \frac{(n-k_1)!}{k_2!(n-k_1-k_2)!} \times \ldots \times \frac{(n-k_1-k_2-\ldots-k_{s-1})!}{(n-k_1-k_2-\ldots-k_{s-1}-k_s)!k_s!}$$

$$= \frac{n!}{k_1!k_2!\ldots k_s!}$$

after cancelling and remembering that $n - k_1 - k_2 - \ldots - k_{s-1} - k_s = 0$.

1.4.1. (a) The numbers 5, 4, 3, 1 can be assigned to the four suits in $4! = 24$ ways. Hence the total number of hands with a 5-4-3-1 suit distribution is

$$24 \times C(13, 5) \times C(13, 4) \times C(13, 3) \times C(13, 1)$$
$$= 24 \times 1287 \times 715 \times 286 \times 13$$
$$= 82\,111\,732\,560.$$

(b) $12 \times C(13,4)^2 \times C(13,3) \times C(13,2) = 136\,852\,887\,600$. (This is 21.6% of all bridge hands, and is the most common of all the suit distributions.)

(c) $4 \times (13,4)^3 \times C(13,1) = 19\,007\,345\,500$.

1.4.2. We assume that given the 26 cards held by you and your partner, any way of giving 13 of the remaining cards to each of your opponents is as likely to occur as any other. Thus the required probability is the ratio $\#(F)/\#(E)$, where E is the set of all the ways your opponent's cards can be divided between them, and F is the subset of those distributions in which each of them has two spades.

We can count the number of hands in E and F by considering just one of your opponents, since, for example, once we know which 13 cards are held by East, it follows that the remaining 13 cards must all be held by West. So we count the hands East might have.

East's hand consists of 13 cards chosen from the 26 cards held by your opponents. So $\#(E) = C(26, 13)$. A hand for East containing exactly two spades is chosen by choosing two of the four spades which your opponents hold and 11 of the 22 non-spades which they have between them. Hence $\#(F) = C(4, 2) \times C(22, 11)$. Hence

$$\frac{\#(F)}{\#(E)} = \frac{C(4, 2) \times C(22, 11)}{C(26, 13)} = \frac{234}{575} = 0.41 \text{ (to 2 decimal places).}$$

1.4.3. We deal with the different types of hand one by one.

(a) Flush. There are $C(13,5) = 1287$ ways of choosing five cards from any one suit. Ten of these will be cards in sequence, thus giving a straight flush. So there are 1277 flushes in each suit, hence $4 \times 1277 = 5108$ flushes altogether.

(b) Four of a kind. The four cards of the same rank can be chosen in 13 ways (as there are 13 ranks altogether). Once these have been chosen, the fifth card in the hand can be any of the other 48 cards left in the pack. So there are $13 \times 48 = 624$ such hands.

(c) Full house. To obtain the full house we first choose three cards of one rank, and then two cards of another rank. The rank of the three cards can be chosen in 13 ways, and then three cards of this rank in $C(4,3) = 4$ ways. Then the rank of the two cards can be chosen in 12 ways, and the two cards of this rank in $C(4,2) = 6$ ways. So the total number of full houses is $13 \times 4 \times 12 \times 6 = 3744$.

(d) One pair. To obtain a one pair hand we first choose the rank of the pair, which we can do in 13 ways. We can then choose two cards of this rank in $C(4,2) = 6$ ways. We then need to choose three different ranks for the remaining three cards, and we can do this in $C(12,3) = 220$ ways. There are four choices of suit for the card of each of these ranks. So the total number of one pair hands is $13 \times 6 \times 220 \times 4^3 = 1\,098\,240$.

(e) Straight. The lowest card in a straight can be any one of the ten cards, A, 2, 3, . . ., 10. We have four choices for the suits of each card in the straight, giving $4^5 = 1024$ combination of suits for the five cards. In four of these cases all five cards will come from the same suit, giving a straight flush. So 1020 of the combinations give straights. So there are $10 \times 1020 = 10\,200$ straights.

(f) Straight flush. The lowest card of a straight flush can be chosen in ten ways, and there are four possible suits, so there are $4 \times 10 = 40$ of these hands.

(g) Three of a kind. The rank of the three cards can be chosen 13 ways, and three cards of this rank in $C(4,3) = 4$ ways. The different ranks of the other two cards can be chosen in $C(12,2) = 66$ ways, and one card of each of these ranks can be chosen in $4 \times 4 = 16$ ways. So there are altogether $13 \times 4 \times 66 \times 16 = 54\,912$ of these hands.

(h) Two pairs. The two ranks of the two pairs can be chosen in $C(13,2) = 78$ ways. For each of these ranks, we can choose two cards of the rank in $C(4,2) = 6$ ways. The rank of the fifth card can then be chosen in 11 ways, and a card of this rank in four ways. Hence there are altogether $78 \times 6 \times 6 \times 11 \times 4 = 123\,552$ of these hands.

(i) Other hands. The only straightforward way to count the number of hands not in any of the above categories is to subtract the total number of the above hands from $C(52, 5) = 2\,598\,960$ which is the total number of poker hands. From the table below we see that this comes to 1 302 540:

Kind of hand	*Number*	*Probability*
Straight Flush	40	0.000015
Four of a Kind	624	0.000240
Full House	3 744	0.001441
Flush	5 108	0.001965
Straight	10 200	0.003925
Three of a Kind	54 912	0.021128
Two Pairs	123 552	0.047539
One Pair	1 098 240	0.422569
Other Hands	1 302 540	0.501177
	2 598 960	

1.4.4. If you exchange three cards, you are picking three cards from the 47 which are not in your hand. This can be done in $C(47, 3) = 16\,215$ ways. We need to count the number of these which lead to a hand which is better than just having one pair. We list the possibilities, and count how many there are of each.

(i) Three cards of the same rank, turning your hand into a full house. Since there are only three 8s among the remaining cards, there is only one way you could choose three 8s, and likewise for three 5s, and three 4s. There are four cards remaining of each of the other nine ranks, and so for each of them three cards can be chosen in $C(4, 3) = 4$ ways. So altogether you could get three cards of the same rank in $(3 \times 1 + 9 \times 4) = 39$ ways.

(ii) Two Jacks and one other card, ending up with four of a kind. The two Jacks can be chosen in just one way since there are just two Jacks left in the pack. The remaining card can then be chosen from the 45 other cards. So this can be done in 45 ways.

(iii) One Jack and a pair of cards of the same rank, again producing a full house. The Jack can be chosen in two ways. For ranks 8, 5 and 4, a pair can be chosen in $C(3, 2) = 3$

ways, and for the other nine ranks in $C(4,2) = 6$ ways. So the total number of ways of choosing a Jack and a pair is $2 \times (3 \times 3 + 9 \times 6) = 126$ ways.

(iv) One Jack and two cards of different ranks, producing three of a kind. The Jack can be chosen in two ways, there are then altogether $C(45,2) = 990$ ways of choosing two from the remaining 45 cards. We saw in case (iii) that 63 of these yield a pair, and hence $990 - 63 = 927$ yield cards of different ranks. Hence there are $2 \times 927 = 1854$ ways of choosing one Jack and two cards of different ranks.

(v) One Pair and a card of a third rank, producing two pairs. As in case (ii) a pair of 8s, 5s or 4s can be chosen in 3×3 ways, and in each of these cases there are $45 - 3 = 42$ possible choices for the remaining card. A pair of cards of each of the other nine ranks can be chosen in six ways, and in each case the remaining card in $45 - 4 = 41$ ways. Hence there are $(3 \times 3 \times 42 + 9 \times 6 \times 41) = 2592$ ways to obtain two pairs in this way.

Next we consider the possibilities that arise if you decide to keep both Jacks and the 8◇. In this case you will draw just two cards from the remaining 47 and this can be done in $C(47,2) = 1081$ ways. Again, we count the number of these which lead to improvements.

(i) Two Jacks, yielding four of a kind. This can happen in just one way.

(ii) Two 8s, yielding a full house. This can happen in $C(3,2) =$ three ways.

(iii) One Jack and one 8, producing a full house again. This can happen in $2 \times 3 =$ six ways.

(iv) One Jack and one card which is neither a Jack nor an 8, producing three of a kind. The Jack can be chosen two ways and then the remaining card in 42 ways, giving 84 cases altogether.

(v) One 8 and a card which is neither a Jack nor an 8, producing two pairs. This can be done in $3 \times 42 = 126$ ways.

(vi) A pair of a rank different from Jacks and 8s, thus again ending up with two pairs. A pairs of 5s can be chosen in three ways, and likewise for a pair of 4s. Pairs of each of the other nine ranks can be chosen in six ways each. So there are $(2 \times 3 + 9 \times 6) = 60$ ways to obtain two pairs in this way.

We can sum up these two cases in the following table (where the percentages are given to 2 decimal places):

	Keeping J♠, J♣		Keeping J♠, J♣, 8♢	
	Number of hands	*% of all choices*	*Number of hands*	*% of all choices*
Four of a kind	45	0.28	1	0.09
Full house	165	1.02	9	0.83
Three of a kind	1854	11.43	84	7.77
Two pairs	2592	15.99	186	17.21
Any improvement	4656	28.71	280	25.90

It can be seen that, although there is not a great deal in it, keeping just the pair of Jacks gives you a better chance of improving your hand.

1.4.5. With 100 balls in the bag there are $C(100, 10)$ ways of selecting a sample of ten balls. Five red and five blue balls can be chosen in $C(50, 5) \times C(50, 5)$ ways. So the required probability is

$$\frac{C(50, 5)^2}{C(100, 10)} = \frac{2\,118\,760}{8\,170\,019} = 0.259 \text{ (to 3 decimal places).}$$

This example shows that the odds are about 3 to 1 against the sample reflecting accurately the proportion of red and blue balls in a bag. If we think of the balls as people whose political opinions are being sampled, it follows that, even assuming that people are being asked sensible questions to which they are giving honest answers, it is unlikely that the results of an opinion poll reflect absolutely accurately the opinions of the electorate. Of course, providing that the sampling has been done carefully, they are unlikely to be very far out, and there is a whole branch of statistics which deals with the question of how accurate the results are likely to be. The upshot of this theory is that for the sample sizes used for most national opinion polls, we can be 90% confident that the results are accurate to within 2–3% (this is on the assumption that the sampling has been done carefully). Thus an opinion poll which shows that support for a party has risen by up to 2% is compatible with the hypothesis that support for that party has actually fallen.

1.4.6. Since each person's birthday can be chosen in 365 ways, when there are n people, their birthdays can be chosen in 365^n ways, that is $\#(E) = 365^n$. It is easier to calculate $\#(E \backslash E_1)$, and hence $\#(E_1)$, rather than calculate $\#(E_1)$ directly, $\#(E \backslash E_1)$ is the number of ways in which n people can have different birthdays. Of course, this is 0 for $n > 365$. For $n \leqslant 365$, $\#(E \backslash E_1)$ is the number of ways of choosing n birthdays in order to assign to the n people. Thus $\#(E \backslash E_1) = P(365, n)$. Thus

$$\frac{\#(E_1)}{\#(E)} = \frac{\#(E) - \#(E \backslash E_1)}{\#(E)}$$

$$= 1 - \frac{\#(E \backslash E_1)}{\#(E)}$$

$$= 1 - \frac{P(365, n)}{365^n}$$

$$= 1 - \frac{365}{365} \times \frac{364}{365} \times \ldots \times \frac{365 - (n-1)}{365}.$$

The smallest integer, n, for which this probability is greater than 0.5, is $n = 23$. Thus with 23 people in a room, there is a better than even chance that two of the people have the same birthday.

1.4.7. Since the sampling is done with replacement, there are $a \times a = a^2$ ways of choosing a red ball twice, and likewise b^2 ways of choosing a blue ball twice. So that total number of ways of choosing two balls of the same colour is

$$a^2 + b^2 = a^2 + (2n - a)^2 = 2((a - n)^2 + n^2),$$

which, with n fixed, clearly attains a minimum when $(a - n)^2$ is as small as possible, that is, when $a = n$, hence when $a = b = n$.

1.4.8. We count the number of ways of allotting k people to n possible birthdays, so that no birthday is shared by more than two people. Let this number be $Q(k, n)$. The total number of ways of assigning birthdays to k people, when there are n dates to choose from, is n^k. So, assuming all n birthdays are equally likely, the probability that no birthday is shared by more than two people is

$$\frac{Q(k, n)}{n^k}.$$

We are interested in the case $n = 365$, and we seek the smallest k such that

$$\frac{Q(k, 365)}{365^k} < \frac{1}{2}. \tag{S1.1}$$

Consider the number of ways of allotting k people to n birthdays so that exactly s birthdays are shared by two people, but no birthday is shared by more than two people. We use $R(k, n, s)$ to stand for this number.

To allot the people to the birthdays in this way, we first choose the s birthdays which are to be shared. We can do this in $C(n, s)$ ways. We then choose successively s pairs of people to be assigned these birthdays. We can do this in

$$C(k, 2)C(k-2, 2) \ldots C(k-2(s-1)), 2) = \frac{k!}{2^s(k-2s)!}$$

ways. We then need to choose $k-2s$ of the remaining $n-s$ birthdays, which we can do in $C(n-s, k-2s)$ ways, and then assign the remaining $k-2s$ people to these birthdays, which we can do in $(k-2s)!$ ways. Hence

$$\begin{aligned} R(k, n, s) &= C(n, s) \times \frac{k!}{2^s(k-2s)!} \\ &\quad \times C(n-s, k-2s) \times (k-2s)! \\ &= \frac{n!k!}{s!2^s(k-2s)!(n-k+s)!}. \end{aligned} \tag{S1.2}$$

Now s can take all values from 0 to $\text{int}(k/2)$. It follows that

$$Q(k, n) = \sum_{s=0}^{\text{int}(k/2)} R(k, n, s). \tag{S1.3}$$

To complete the problem we need to solve inequality (S1.1) using the formula for $Q(k, 365)$ given by equations (S1.2) and (S1.3). Not an easy sum to do by hand! But using a computer it turns out that the least k which satisfies (S1.1) for $n = 365$, is $k = 88$. Hence with 88 people in a room there is a better than even chance that there are three people who share a birthday.

I set this problem in the *Mathematical Gazette* in 1986. Several people supplied a solution, one of whom, K. Thomas, provided a particularly elegant method for doing the necessary calculations. Professor C. A. Rankin and A. C. Robin came up with 187 for the number of people required before there is a better than even chance that four people share a birthday.*

If you cannot assemble groups of 88 people to test this answer empirically, you can try it out with any reference book that gives birthdays. Open the book at random, start with a particular person, and see how many people you have to check before you come across three people who share a birthday. When I tried this out with *The Fontana Biographical Companion to Modern Thought*, edited by Alan Bullock and R. B. Woodings (London, 1983), starting at the beginning, the first triple coincidence of birthdays that I found involved the Roumanian poet Tudor

* *Mathematical Gazette*, 70 (1986), p. 52 and pp. 228–9.

Arghezi, the German Jewish religious leader Leo Baeck and the American physicist John Bardeen, who were all born on 23 May, in 1880, 1873 and 1908, respectively. Not counting people whose birthdays are not given, John Bardeen is the 87th person listed.

CHAPTER 2

2.1.1. We let P_k be the set of integers in the range from 1 to 1 000 000 which are perfect kth powers. We want $\#(P_2 \cup P_3)$. By the Inclusion–Exclusion Theorem, this is given by

$$\#(P_2 \cup P_3) = \#(P_2) + \#(P_3) - \#(P_2 \cap P_3).$$

A number is in $P_2 \cap P_3$ if and only if it is an integer in the range from 1 to 1 000 000 which is both a perfect square and a perfect cube, that is, if and only if it is an integer in this range which is a perfect sixth power. Since $1\,000\,000 = 10^6$, it follows that this set contains ten numbers. Thus, we have

$$\#(P_2 \cup P_3) = 1\,000 + 100 - 10 = 1090.$$

2.1.2. We let D_k be the set of integers in the range from 1 to 1 000 000 which are divisible by k. By the Inclusion–Exclusion Theorem,

$$\begin{aligned}
&\#(D_2 \cup D_3 \cup D_5 \cup D_7) \\
&= \big(\#(D_2) + \#(D_3) + \#(D_5) + \#(D_7)\big) - \big(\#(D_2 \cap D_3) \\
&\quad + \#(D_2 \cap D_5) + \#(D_2 \cap D_7) + \#(D_3 \cap D_5) + \#(D_3 \cap D_7) \\
&\quad + \#(D_5 \cap D_7)\big) + \big(\#(D_2 \cap D_3 \cap D_5) + \#(D_2 \cap D_3 \cap D_7) \\
&\quad + \#(D_2 \cap D_5 \cap D_7) + \#(D_3 \cap D_5 \cap D_7)\big) \\
&\quad - \#(D_2 \cap D_3 \cap D_5 \cap D_7) \\
&= \big(\#(D_2) + \#(D_3) + \#(D_5) + \#(D_7)\big) \\
&\quad - \big(\#(D_6) + \#(D_{10}) + \#(D_{14}) + \#(D_{15}) + \#(D_{21}) + \#(D_{35})\big) \\
&\quad + \big(\#(D_{30}) + \#(D_{42}) + \#(D_{70}) + \#(D_{105})\big) - \#(D_{210}) \\
&= (500\,000 + 333\,333 + 200\,000 + 142\,857)) \\
&\quad - (166\,666 + 100\,000 + 71\,428 + 66\,666 + 47\,619 + 28\,571) \\
&\quad + (33\,333 + 23\,809 + 14\,285 + 9\,523) - 4\,761 \\
&= 771\,429.
\end{aligned}$$

It follows that the number of integers in the range from 1 to 1 000 000 which are *not* divisible by any of 2, 3, 5, 7 is

$$1\,000\,000 - 771\,429 = 228\,571.$$

2.1.3. A positive integer less than or equal to n has a prime factor in common with n if and only if it is divisible by one of $p_1, \ldots, p_k$. Let E_i be the set of integers in the range from 1 to n which are divisible by p_i. By the Inclusion–Exclusion Theorem,

$$\#\left(\bigcup_{i=1}^{k} E_i\right) = \sum_{1 \leqslant i \leqslant k} \#(E_i) - \sum_{1 \leqslant i_1 < i_2 \leqslant k} \#(E_{i_1} \cap E_{i_2}) + \ldots$$

$$= \sum_{1 \leqslant i \leqslant k} \frac{n}{p_i} - \sum_{1 \leqslant i_1 < i_2 \leqslant k} \frac{n}{p_{i_1} p_{i_2}} + \ldots$$

Hence the number of integers in the range from 1 to n which have no prime factors in common with n is

$$n - \left(\sum_{1 \leqslant i \leqslant k} \frac{n}{p_i} - \sum_{1 \leqslant i_1 < i_2 \leqslant k} \frac{n}{p_{i_1} p_{i_2}} + \ldots\right)$$

$$= n\left(1 - \sum_{1 \leqslant i \leqslant k} \frac{1}{p_i} - \sum_{1 \leqslant i_1 < i_2 \leqslant k} \frac{1}{p_{i_1} p_{i_2}} + \ldots\right)$$

Hence, using the algebraic identity,

$$(1 - x_1)(1 - x_2) \ldots (1 - x_k)$$

$$= 1 - \sum_{1 \leqslant i \leqslant k} x_i + \sum_{1 \leqslant i_1 < i_2 \leqslant k} x_{i_1} x_{i_2} + \ldots,$$

it follows that

$$\varphi(n) = n\left(1 - \frac{1}{p_1}\right)\left(1 - \frac{1}{p_2}\right) \ldots \left(1 - \frac{1}{p_k}\right).$$

2.1.4. Throwing a dice amounts to sampling with replacement the six numbers on the faces of the dice. Hence this question is essentially the same as that of Problem 2 in section 2.1, with $n = 6$. Thus the probability that each of the numbers from 1 to 6 is thrown at least once in a sequence of s throws is

$$\theta_6(s) = \frac{1}{6^s} \sum_{k=0}^{6} (-1)^k C(6, k)(6 - k)^s.$$

Hence

$$\theta_6(s) = 1 - 6(\tfrac{5}{6})^s + 15(\tfrac{4}{6})^s - 20(\tfrac{3}{6})^s + 15(\tfrac{2}{6})^s - 6(\tfrac{1}{6})^s.$$

We want the least s such that $\theta_6(s) > 0.5$. This can be found by trying values of s in the formula above. You will find that $\theta_6(12) = 0.438$ and $\theta_6(13) = 0.514$, and hence the required value of s turns out to be $s = 13$.

2.2.1. There are altogether 10! ways of putting ten letters in ten envelopes. The number of derangements, that is, cases where every letter is in the wrong envelope, is given by Theorem 2.5, with $n = 10$. The required probability is this number divided by 10!, that is,

$$\sum_{k=0}^{10} \frac{(-1)^k}{k!} = 0.368 \text{ (to three decimal places).}$$

2.2.2. Take the solution to Problem 3, substituting n for 52.

2.2.3. (i) There are two anagrams in which the D is replaced by an A, since in this case the As must be replaced by the Rs, leaving two possible arrangements for the remaining A and the D. Likewise there are two anagrams in which the D is replaced by an R. So there are altogether 4 anagrams: DRARA, ARARD, ARRDA, ADRRA.

(ii) The As can replace any three of the letters N, G, R, M, and this can be done in four ways. The letter not replaced by an A must itself replace an A, which can be done in three ways, and the remaining letters can fill the remaining three places in any order, which can be done in 3! ways. So the number of anagrams is $4 \times 3 \times 3! = 72$.

CHAPTER 3

3.1.1. $p(1) = 1$; $p(2) = 2$; $p(3) = 3$; $p(4) = 5$;
$p(5) = 7$; $p(6) = 11$; $p(7) = 15$; $p(8) = 22$.

3.1.2. $p_3(1) = 1$; $p_3(2) = 2$; $p_3(3) = 3$; $p_3(4) = 4$; $p_3(5) = 5$;
$p_3(6) = 7$; $p_3(7) = 8$; $p_3(8) = 10$; $p_3(9) = 12$; $p_3(10) = 14$.

3.1.3. Partitioning n into an ordered sum is equivalent to splitting up a row of n objects into blocks. For example, corresponding to the ordered sum

$$2 + 3 + 1 + 1 + 5 = 12$$

we have 4 markers splitting up a row of 12 dots as follows:

$$\cdot\ \cdot | \cdot\ \cdot\ \cdot | \cdot | \cdot | \cdot\ \cdot\ \cdot\ \cdot\ \cdot$$

With a row of n dots there are $n - 1$ spaces between the dots, and for each space we have two choices, either to split up the row with a marker in that space, or not. This gives 2^{n-1} choices altogether, and thus this is the number of ways of writing n as an ordered sum.

3.2.1. The required formula is

$$p_3(n) = \mathrm{int}\left(\frac{n^2}{12} + \frac{n}{2} + 1\right). \qquad \text{(S3.1)}$$

In fact,

$$p_3(n) = \frac{n^2}{12} + \frac{n}{2} + \delta_n, \qquad \text{(S3.2)}$$

where

$$\delta_n = \begin{cases} 1 & \text{if } n \equiv 0 \pmod 6, \\ \frac{5}{12} & \text{if } n \equiv 1 \pmod 6 \text{ or } n \equiv 5 \pmod 6, \\ \frac{2}{3} & \text{if } n \equiv 2 \pmod 6 \text{ or } n \equiv 4 \pmod 6, \\ \frac{3}{4} & \text{if } n \equiv 3 \pmod 6. \end{cases}$$

It is easily seen that equation (S3.1) is an immediate consequence of equation (S3.2).

These formulas can be discovered in various ways. Once discovered, they are very easy to prove by mathematical induction as follows:

It is straightforward to check that for $n = 1, 2, 3, 4, 5, 6$ these formulas do fit the values of p_3, as given in the answer to Exercise 3.1.2.

Now suppose that $m > 6$, and that equation (S3.2) holds for all positive integers less than m. By Theorem 3.4,

$$\begin{aligned} p_3(m) &= p_2(m) + p_3(m-3) \\ &= p_2(m) + p_2(m-3) + p_3(m-6) \\ &= \mathrm{int}\left(\frac{m}{2} + 1\right) + \mathrm{int}\left(\frac{m-3}{2} + 1\right) + p_3(m-6), \qquad \text{(S3.3)} \end{aligned}$$

using Lemma 3.1(c). Now one of $m, m-3$ will be even and the other odd. So in taking the integer parts in equation (S3.3), in one case we get the number itself and in the other case we subtract $\frac{1}{2}$. Thus it follows from equation (S3.3) that

$$\begin{aligned} p_3(m) &= \left(\frac{m}{2} + 1\right) + \left(\frac{m-3}{2} + 1\right) - \frac{1}{2} + p_3(m-6) \\ &= p_3(m-6) + m. \end{aligned}$$

By our induction hypothesis, formula (S3.2) holds when n is replaced by $m - 6$, and hence

$$p_3(m) = \frac{(m-6)^2}{12} + \frac{(m-6)}{2} + \delta_{m-6} + m$$
$$= \frac{m^2}{12} + \frac{m}{2} + \delta_m,$$

as $\delta_{m-6} = \delta_m$. Thus formula (S3.2) holds also for $n = m$. Hence, by Mathematical Induction, formula (S3.2) holds for all $n \in \mathbb{N}^+$.

3.2.2. (a) $s_1(10) = 30$; $s_2(10) = 7$; $s_3(10) = 2$;
$s_4(10) = 1$; $s_5(10) = 1$.

(b) We divide up the partitions of n according to the size of the smallest term. n has one partition consisting just of one term, namely n itself. Otherwise a partition of n contains at least two terms, and hence the smallest term cannot be larger than $n/2$. So the number of partitions of n equals the sum of the numbers of partitions of n with smallest term t, for $1 \leqslant t \leqslant \text{int}(n/2)$, plus 1.

(c) We divide up the partitions of n with smallest term k according to the size of the smallest term when we remove one term of size k. When we do this we are left with a partition of $n - k$ with smallest term t, where $k \leqslant t \leqslant n - k$.

(d) If we add a term of size 1 to a partition of n we obtain a partition of $n + 1$ whose smallest term is 1. Conversely, given a partition of $n + 1$ with smallest term 1, by removing this term we get a partition of n. This defines a one-to-one correspondence between the partitions of n and the partitions of $n + 1$ with smallest term 1. Hence $p(n) = s_1(n + 1)$.

(e) The formula can be derived from part (c), but we prefer a direct combinatorial argument. The partitions of n with smallest term k fall into disjoint classes.

The first class consists of those partitions which have at least two terms of size k. Removing one of these terms of size k, we get a partition of $n - k$, also with smallest term k, and conversely. Hence there are $s_k(n - k)$ partitions in this class.

The second class consists of those partitions which have only one term of size k. Hence if we add 1 to this term we get a partition of $n + 1$ with smallest term $k + 1$. Conversely, given such a partition of $n + 1$, by removing 1 from the smallest term, we get a partition of n with smallest term k which has just one term of this size. Hence there are $s_{k+1}(n + 1)$ partitions in this class.

We have thus counted all the partitions of n with smallest term k, and the formula follows immediately.

3.3.1. The answer depends, of course, on how the calculation is done. Given n it takes one multiplication to calculate n^2, a further multiplication to calculate n^3, and so on. Thus we need $k-1$ multiplications to obtain the values of $n, n^2, \ldots, n^k$. We need a further k multiplications to work out the products $a_i n^i$, for $1 \leqslant i \leqslant k$, and then k additions to evaluate $f(n)$. Thus, with this approach we need $2k-1$ multiplications and k additions.

However, if we rewrite the formula for $f(n)$ as

$$f(n) = a_0 + n(a_1 + n(a_2 + n(a_3 + \ldots + n(a_{k-1} + na_k) \ldots))),$$

then we see that we can calculate $f(n)$ with just k multiplications and k additions.

Either way, we see that the number of arithmetic operations needed to evaluate $f(n)$ does not increase as n increases. Of course, as n increases, it will take longer and longer to do the multiplications by n. None the less, the values of a polynomial function can be calculated very efficiently. [Contrast this with the problem of evaluating $n!$. It takes $n-2$ multiplications to evaluate the product $1 \times 2 \times 3 \times \ldots \times n$ (there are $n-1$ multiplication signs here, but multiplication by 1 involves no arithmetic). So not only do the individual multiplications take longer as n increases, the number of multiplications that we have to do also increases.]

3.3.2. The following program reads in the value of n, enumerates the partitions of n in a two-dimensional array $A(i, j)$, and then prints out the number of these partitions. Since the terms in one partition are calculated from the terms in the previous partition we save storage space by not storing all the previous partitions that we have calculated. Instead, i oscillates between the values 1 and 2, and we use an extra variable, COUNT, to keep track of how many partitions we have found.

```
      INTEGER A,COUNT,I,I1,J,J1,J2,M,N,SUM
      DIMENSION A(2,50)
      WRITE(6,10)
   10 FORMAT('WHAT IS THE VALUE OF N?')
      READ(5,*) N
      COUNT=1
      I=1
      A(1,1)=N
      A(1,2)=0
   20 IF(A(I,1).EQ.1) GO TO 100
      SUM=0
      I1=I
```

```
      I=3-I
      COUNT=COUNT+1
      J=1
 30   IF(A(I1,J).GT.1) THEN
        J=J+1
        GO TO 30
      ELSE
        J=J-1
        IF(J.EQ.1) THEN
              A(I,1)=A(I1,1)-1
              SUM=A(I,1)
              GO TO 60
        ELSE
              J1=J-1
              GO TO 40
        ENDIF
      ENDIF
 40   DO 50 J=1,J1
      A(I,J)=A(I1,J)
 50   SUM=SUM+A(I,J)
      J=J1+1
      A(I,J)=A(I1,J)-1
      SUM=SUM+A(I,J)
 60   M=N-SUM
      J2=J
      J=J+1
      IF (M.GT.A(I,J2)) THEN
        A(I,J)=A(I,J2)
        SUM=SUM+A(I,J)
        GO TO 60
      ELSE
        A(I,J)=M
        J=J+1
        A(I,J)=0
        GO TO 20
      ENDIF
100   WRITE(6,110) COUNT,N
110   FORMAT('THERE ARE',I6,' PARTITIONS OF',I3)
      STOP
      END
```

You will be able to deduce the age of the author from the fact that this program is written in FORTRAN, which many people will regard as rather old-fashioned. It is hoped that you will find it straightforward to translate this program into your favourite language.

The program does not display nor print out the partitions as they are enumerated. It would be very straightforward to add some code to do this but this would slow down the running of the program. It would also take up a lot of paper to print all the partitions once n gets much above 20.

Note that the dimension statement in this program specifies that A is a 2×50 array. The 2 is to allow i to oscillate between the values 1 and 2. The 50 is the largest number of terms that can arise in any partition. Since the largest number of terms that arise in a partition of n is n itself, this dimension number would need to be increased if this program is to calculate $p(n)$ for $n > 50$.

CHAPTER 4

4.1.1. By stipulation, we have ensured $\sim$ is a reflexive relation. Suppose now that $f \sim g$. Then

$$\lim_{n\to\infty} \frac{f(n)}{g(n)} = 1.$$

Hence, by the Reciprocal Rule for limits,

$$\lim_{n\to\infty} \frac{g(n)}{f(n)} = \frac{1}{\lim_{n\to\infty} \frac{f(n)}{g(n)}} = \frac{1}{1} = 1$$

and hence $g \sim f$. Thus $\sim$ is symmetric. Finally suppose that $f \sim g$ and $g \sim h$. Then

$$\lim_{n\to\infty} \frac{f(n)}{h(n)} = \lim_{n\to\infty} \left(\frac{f(n)}{g(n)} \times \frac{g(n)}{h(n)}\right)$$

$$= \lim_{n\to\infty} \frac{f(n)}{g(n)} \times \lim_{n\to\infty} \frac{g(n)}{h(n)},$$

by the Product Rule for limits,

$$= 1 \times 1 = 1.$$

Hence $f \sim h$. This shows that $\sim$ is transitive, thus completing the proof.

4.1.2. Suppose that $f \sim g$. Then the sequence $\{f(n)/g(n)\}$ has a limit and so is bounded. Hence there is some constant, C, such that for all $n \in \mathbb{N}$,

$$\left| \frac{f(n)}{g(n)} \right| < C,$$

and hence

$$|f(n)| < C|g(n)|.$$

4.1.3. If $a_k \neq 0$, then

$$\lim_{n\to\infty} \frac{a_k n^k + \ldots + a_1 n + a_0}{a_k n^k} = \lim_{n\to\infty} \frac{a_k + \ldots + a_1/n^{k-1} + a_0/n^k}{a_k}$$

$$= \frac{a_k + \ldots + 0 + 0}{a_k} = 1,$$

using the Sum and Quotient Rules for limits. This proves that

$$a_k n^k + \ldots + a_1 n + a_0 \sim a_k n^k.$$

Suppose now that $j \neq k$, say $j > k$. Then

$$\lim_{n\to\infty} \frac{b_k n^k}{a_j n^j} = \lim_{n\to\infty} \frac{b_k}{a_j n^{j-k}} = 0$$

and thus $a_j n^j$ is not asymptotic to $b_k n^k$. Conversely, if $j = k$, then

$$\lim_{n\to\infty} \frac{b_k n^k}{a_j n^j} = \frac{b_k}{a_j}$$

and so equals 1 if and only if $a_j = b_k$.

4.1.4. Let $A = \max\{f(n) : 1 \leqslant t < n_0\}$, and let h be the polynomial function, $h : n \mapsto g(n) + A$. Then for all $n \in \mathbb{N}$, $f(n) \leqslant h(n)$.

4.2.1. Integrating by parts, we obtain

$$\int_0^{\pi/2} \sin^n x \, dx = \int_0^{\pi/2} \sin x \sin^{n-1} x \, dx$$

$$= [(-\cos x) \sin^{n-1} x]_0^{\pi/2}$$

$$- \int_0^{\pi/2} (-\cos x)(n-1) \sin^{n-2} x \cos x \, dx$$

$$= 0 + (n-1) \int_0^{\pi/2} (1 - \sin^2 x) \sin^{n-2} x \, dx$$

$$= (n-1)(I_{n-2} - I_n).$$

It follows that $nI_n = (n-1)I_{n-1}$, whence

$$I_n = \frac{n-1}{n} I_{n-1}. \tag{S4.1}$$

Now $I_0 = \int_0^{\pi/2} 1 \, dx = \pi/2$, and hence the formula

$$I_{2n} = \frac{(2n)!\pi}{2^{2n+1}(n!)^2} \tag{S4.2}$$

holds for $n = 0$. Suppose now that it holds for $n = m$. Then, we have, by equation (S4.1)

$$\begin{aligned} I_{2(m+1)} &= \frac{2m+1}{2m+2} I_{2m} \\ &= \frac{2m+1}{2m+2} \cdot \frac{(2m)!\pi}{2^{2m+1}(m!)^2} \\ &= \frac{2m+2}{2m+2} \cdot \frac{2m+1}{2m+2} \cdot \frac{(2m)!\pi}{2^{2m+1}(m!)^2} \\ &= \frac{(2m+2)!\pi}{2^{2m+3}((m+1)!)^2}, \end{aligned}$$

which shows that equation (S4.2) holds also for $n = m + 1$. It follows by Mathematical Induction that equation (S4.2) holds for all $n \in \mathbb{N}$. The proof of the corresponding formula for I_{2n+1} is similar.

4.2.2. We replace $n!$ by the approximation given by Stirling's Theorem, and solve the inequality,

$$\left(\frac{n}{\mathrm{e}}\right)^n \sqrt{2\pi n} > 10^{10^6}.$$

Taking logarithms to base 10, this is equivalent to

$$n \log_{10}\left(\frac{n}{\mathrm{e}}\right) + \frac{1}{2}\log_{10}(2\pi n) > 10^6.$$

By trial and error, we find that the smallest integer solution of this inequality is $n = 205\,023$. Stirling's approximation is accurate enough for this also to be the smallest integer solution of

$$n! > 10^{10^6}.$$

4.2.3. If a coin is tossed $2n$ times there are 2^{2n} sequences of possible outcomes. The number of these sequences with an equal number of heads and tails is equal to the number of ways of choosing which n tosses in the sequence are to be heads, that is $C(2n, n)$.

Hence the required probability is

$$\frac{C(2n, n)}{2^{2n}} = \frac{(2n)!}{2^{2n}(n!)^2}.$$

By Stirling's Formula, $n!$ is asymptotic to

$$\left(\frac{n}{\mathrm{e}}\right)^n \sqrt{2\pi n}\,.$$

Hence $(n!)^2$ is asymptotic to

$$\left(\left(\frac{n}{\mathrm{e}}\right)^n \sqrt{2\pi n}\right)^2 = \left(\frac{n}{\mathrm{e}}\right)^{2n} 2\pi n$$

and $(2n)!$ is asymptotic to

$$\left(\frac{2n}{\mathrm{e}}\right)^{2n} \sqrt{4\pi n}\,.$$

It follows that

$$\frac{C(2n,\, n)}{2^{2n}}$$

is asymptotic to

$$\frac{\left(\frac{2n}{\mathrm{e}}\right)^{2n} \sqrt{4\pi n}}{2^{2n}\left(\frac{n}{\mathrm{c}}\right)^{2n} 2\pi n}.$$

Most of the terms cancel and we see that this expression equals

$$\frac{1}{\sqrt{\pi n}}.$$

4.3.1. Note, that to differentiate the function

$$x \mapsto \frac{n^x \mathrm{e}^{2x}}{x^{2x}},$$

it simplifies matters to rewrite this function in the equivalent form

$$x \mapsto \mathrm{e}^{x\ln(n)+2x-2x\ln(x)},$$

from which we can easily see that the derivative is the function

$$x \mapsto (\ln(n) - 2\ln(x))\mathrm{e}^{x\ln(n)+2x-2x\ln(x)}$$

and the second derivative is

$$x \mapsto \left((\ln(n) - 2\ln(x))^2 - \frac{2}{x}\right) \mathrm{e}^{x\ln(n)+2x-2x\ln(x)}$$

We see that the first derivative is 0 when $\ln(n) = 2\ln(x)$, that is, just when $x = \sqrt{n}$. For this value of x, we see that the second derivative is negative, showing that $x = \sqrt{n}$ does give a maximum.

4.3.2. We give the values, correct to three decimal places, in the following table:

n	$p(n)$	$\frac{1}{2\pi(1.1)^2 e^2}\left(\frac{e^{2\sqrt{n}}}{n}\right)$	$\frac{1}{4\sqrt{3}}\left(\frac{e^{\pi\sqrt{[(2/3)n]}}}{n}\right)$
1	1	0.132	1.877
2	2	0.151	2.715
3	3	0.190	4.091
4	5	0.243	6.100
5	7	0.312	8.941
6	11	0.398	12.882
7	15	0.505	18.267
8	22	0.637	25.538
9	30	0.798	35.250
10	42	0.993	48.104

It can be seen that our lower bound given in inequality (4.22) grossly underestimates the values of $p(n)$ in this range, whereas the Hardy–Ramanujan asymptotic formula gives quite a good approximation to the values of $p(n)$.

CHAPTER 5

5.2.1. (a) We know that

$$\frac{1}{(1-x)^2} = \sum_{n=0}^{\infty} nx^{n-1}.$$

Differentiating both sides of this equation we obtain

$$\frac{2}{(1-x)^3} = \sum_{n=0}^{\infty} n(n-1)x^{n-2},$$

and hence on multiplying both sides by x^2, we see that the generating function for the sequence $\{n(n-1)\}$ is

$$x \mapsto \frac{2x^2}{(1-x)^3}.$$

(b) We know, from the generating function for the sequence $\{n\}$ that

$$\frac{x}{(1-x)^2} = \sum_{n=0}^{\infty} nx^n.$$

Differentiating both sides of this equation we obtain

$$\frac{(1+x)}{(1-x)^3} = \sum_{n=0}^{\infty} n^2x^{n-1},$$

and hence, on multiplying through by x, we see that the generating function for the sequence $\{n^2\}$ is

$$x \mapsto \frac{x(1+x)}{(1-x)^3}.$$

(c) Likewise we obtain the generating function for the sequence $\{n^3\}$ by taking the generating function for the sequence n^2, differentiating, and multiplying by x. In this way we obtain

$$x \mapsto \frac{x(1+4x+x^2)}{(1-x)^4}$$

as the generating function for the sequence $\{n^3\}$.

5.3.1. To obtain the generating function for the sequence $\{o(n)\}$, we drop from the infinite product for the generating function, P, for the sequence $\{p(n)\}$, those terms corresponding to parts of even size. Thus the generating function, O, is given by

$$O(x) = \frac{1}{(1-x)(1-x^3)(1-x^5)\dots}.$$

We have seen that the generating function for the sequence $\{q^{\neq}(n)\}$ is given by

$$\begin{aligned} Q^{\neq}(x) &= (1+x)(1+x^2)(1+x^3)\dots \\ &= \frac{(1-x^2)}{(1-x)}\cdot\frac{(1-x^4)}{(1-x^2)}\cdot\frac{(1-x^6)}{(1-x^3)}\dots \end{aligned}$$

In this infinite product all the terms in the numerator cancel the term in the denominator of even degree, leaving just the terms of odd degree. Thus

$$Q^{\neq}(x) = O(x)$$

and hence, by equating coefficients, it follows that for all $n \in \mathbb{N}$,

$$q^{\neq}(n) = o(n).$$

5.3.2. Using the same technique of dropping the appropriate terms from the generating function, P, we see that the generating function for the sequence $\{q^{\nmid 3}(n)\}$ is

$$Q^{\nmid 3}(x) = \frac{1}{(1-x)(1-x^2)(1-x^4)(1-x^5)(1-x^7)\dots}$$

and the generating function for the sequence $\{q^{<3}(n)\}$ is

$$Q^{<3}(x) = (1+x+x^2)(1+x^2+x^4)(1+x^3+x^6)\dots$$

Since, for $x \neq 1$, $(1+x^k+x^{2k}) = (1-x^{3k})/(1-x^k)$, it follows that

$$(1 + x + x^2)(1 + x^2 + x^4)(1 + x^3 + x^6) \dots$$
$$= \frac{(1 - x)^3}{(1 - x)} \cdot \frac{(1 - x)^6}{(1 - x^2)} \cdot \frac{(1 - x)^9}{(1 - x^3)} \cdots$$
$$= Q^{<3}(x),$$

and hence for all $n \in \mathbb{N}$, $q^{\nmid 3}(n) = q^{<3}(n)$.

CHAPTER 6

6.2.1. Let b_n be the coefficient of x^n in

$$\sum_{k=0}^{\infty} (9x(1 + x))^k.$$

(Note that for convenience we have changed the summation variable from n to k.) The $(k + 1)^{\text{th}}$ term in this sum, namely, $(9x(1 + x))^k$ contains terms in x^n for $k \leqslant n \leqslant 2k$, and the coefficient of x^n is $9^k C(k, n - k)$. Hence

$$b_n = \sum_{k \leqslant n \leqslant 2k} 9^k C(k, n - k).$$

The coefficient of x^n in $(1 + x) \sum_{k=0}^{\infty} (9x(1 + x))^k$ is $b_n + b_{n-1}$, that is,

$$a_n = \sum_{k \leqslant n \leqslant 2k} 9^k C(k, n - k) + \sum_{k \leqslant n-1 \leqslant 2k} 9^k C(k, n - 1 - k).$$

6.2.2. (a) Let F be the generating function for the Fibonacci sequence $\{f_n\}$. From the recurrence relation

$$f_{n+1} = f_n + f_{n-1}, \qquad \text{for } n \geqslant 2,$$

it follows that

$$\sum_{n=2}^{\infty} f_{n+1} x^{n+1} = x \sum_{n=2}^{\infty} f_n x^n + x^2 \sum_{n=2}^{\infty} f_{n-1} x^{n-1}.$$

Hence, as $f_1 = f_2 = 1$,

$$F(x) - x - x^2 = x(F(x) - x) + x^2 F(x),$$

from which it follows that

$$F(x) = \frac{x}{(1 - x - x^2)}.$$

This gives the generating function for the Fibonacci sequence.

(b) $(1 - x - x^2) = (1 - \alpha x)(1 - \beta x)$, where $\alpha = (1 + \sqrt{5})/2$ and $\beta = (1 - \sqrt{5})/2$. $(1 - x - x^2 = 0 \Leftrightarrow x = 1/\alpha$ or $x = 1/\beta$. So

α, β are the reciprocals of the zeros of $1 - x - x^2$.) Using the technique of partial fractions, as described in section 6.6, it follows that

$$\frac{x}{(1 - x - x^2)} = \frac{1}{\alpha - \beta}\left(\frac{1}{1 - \alpha x} - \frac{1}{1 - \beta x}\right)$$

$$= \frac{1}{\alpha - \beta}\left(\sum_{n=0}^{\infty} \alpha^n x^n - \sum_{n=0}^{\infty} \beta^n x^n\right).$$

Equating coefficients, it follows that

$$f_n = \frac{1}{\alpha - \beta}(\alpha^n - \beta^n).$$

(c) Alternatively, we can write the generating function, $F(x)$, as follows:

$$F(x) = \frac{x}{(1 - x - x^2)}$$

$$= \frac{x}{(1 - [x(1 + x)])} = x \sum_{k=0}^{\infty} x^k (1 + x)^k.$$

Now, $x^k(1 + x)^k = x^k \Sigma_{t=0}^{k} C(k, t) x^k$, and therefore includes a term of degree n provided that $k \leqslant n \leqslant 2k$, that is for $\frac{1}{2}n \leqslant k \leqslant n$. When $k \leqslant n \leqslant 2k$ the coefficient of x^n in $x^k \Sigma_{t=0}^{k} C(k, t) x^t$ is $C(k, n - k)$. So the coefficient of x^{n+1} in $F(x)$ is given by

$$f_{n+1} = \sum_{n/2 \leqslant k \leqslant n} C(k, n - k).$$

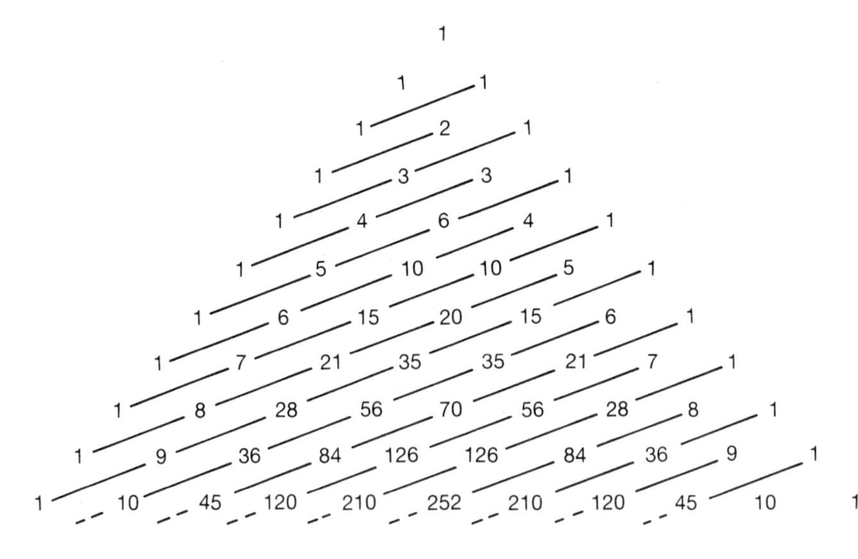

The entries $C(k, n-k)$, for $\frac{1}{2}n \leqslant k \leqslant n$, occur on the diagonals of Pascal's Triangle as shown in the diagram above. It follows from the above formula that the sums of the numbers on these diagonals give the terms of the Fibonacci sequence.

6.3.1. Let $A = x \mapsto \Sigma_{n=1}^{\infty} a_n x^n$ be the generating function for the sequence $\{a_n\}$. From the recurrence relation

$$a_{n+2} = 3a_{n+1} + 4a_n$$

it follows that

$$\frac{1}{x^2}\sum_{n=1}^{\infty} a_{n+2} = \frac{3}{x}\sum_{n=1}^{\infty} a_{n+1}x^{n+1} + 4\sum_{n=1}^{\infty} a_n x^n.$$

That is,

$$\frac{1}{x^2}(A - x - 3x^2) = \frac{3}{x}(A - x) + 4A,$$

from which it follows that

$$A = \frac{x}{(1-4x)(1+x)} = \frac{1}{5}\left(\frac{1}{1-4x} - \frac{1}{1+x}\right)$$
$$= \frac{1}{5}\left(\sum_{n=0}^{\infty} 4^n x^n - \sum_{n=0}^{\infty} (-1)^n x^n\right).$$

Equating coefficients, it follows that

$$a_n = \tfrac{1}{5}(4^n - (-1)^n).$$

6.3.2. Consider a sequence of $n+2$ 0s, 1s and 2s in which 0 can only be followed by a 1. Such sequences have the form $x\$_1$ or $01\$_2$ where x is 1 or 2 and $\$_1$, $\$_2$ are sequences of lengths $n+1$, n respectively, with the required property. It follows that

$$a_{n+2} = 2a_{n+1} + a_n.$$

Also $a_1 = 3$, and $a_2 = 7$. If A is the generating function for the sequence $\{a_n\}$, it follows from the recurrence relation that

$$\frac{1}{x^2}\sum_{n=1}^{\infty} a_{n+2}x^{n+2} = \frac{2}{x}\sum_{n=1}^{\infty} a_{n+1}x^{n+1} + \sum_{n=1}^{\infty} a_n x^n.$$

That is

$$\frac{1}{x^2}(A - 3x - 7x^2) = \frac{2}{x}(A - 3x) + A$$

from which it follows that

$$A = x \mapsto \frac{x^2 + 3x}{(1 - 2x - x^2)}.$$

6.3.3. By expanding the determinant down the first column, we see that

$$\det(A_n) = \begin{vmatrix} 1 & 1 & 0 & 0 & . & 0 & 0 & 0 \\ 1 & 1 & 1 & 0 & . & 0 & 0 & 0 \\ 0 & 1 & 1 & 1 & . & 0 & 0 & 0 \\ . & . & . & . & . & . & . & . \\ . & . & . & . & . & . & . & . \\ 0 & 0 & 0 & 0 & . & 1 & 1 & 0 \\ 0 & 0 & 0 & 0 & . & 1 & 1 & 1 \\ 0 & 0 & 0 & 0 & . & 0 & 1 & 1 \end{vmatrix}$$

$$- \begin{vmatrix} 1 & 0 & 0 & 0 & . & 0 & 0 & 0 \\ 1 & 1 & 1 & 0 & . & 0 & 0 & 0 \\ 0 & 1 & 1 & 1 & . & 0 & 0 & 0 \\ . & . & . & . & . & . & . & . \\ . & . & . & . & . & . & . & . \\ 0 & 0 & 0 & 0 & . & 1 & 1 & 0 \\ 0 & 0 & 0 & 0 & . & 1 & 1 & 1 \\ 0 & 0 & 0 & 0 & . & 0 & 1 & 1 \end{vmatrix}$$

$$= \det(A_{n-1}) - \begin{vmatrix} 1 & 1 & 0 & . & 0 & 0 & 0 \\ 1 & 1 & 1 & . & 0 & 0 & 0 \\ . & . & . & . & . & . & . \\ . & . & . & . & . & . & . \\ 0 & 0 & 0 & . & 1 & 1 & 0 \\ 0 & 0 & 0 & . & 1 & 1 & 1 \\ 0 & 0 & 0 & . & 0 & 1 & 1 \end{vmatrix}$$

$$= \det(A_{n-1}) - \det(A_{n-2}),$$

on expanding the second determinant in the first line along the top row. We therefore see that the required recurrence relation is

$$a_n = a_{n-1} - a_{n-2}.$$

Also, $a_1 = |1| = 1$, and

$$a_2 = \begin{vmatrix} 1 & 1 \\ 1 & 1 \end{vmatrix} = 0.$$

The general solution of this recurrence relation is

$$a_n = A\alpha^n + B\beta^n$$

where α, β are the solutions of

$$x^2 - x + 1 = 0.$$

Thus

$$\alpha = \tfrac{1}{2}(1 + \sqrt{3}\mathrm{i}) = \mathrm{e}^{\mathrm{i}\pi/3} \quad \text{and} \quad \beta = \tfrac{1}{2}(1 - \sqrt{3}\mathrm{i}) = \mathrm{e}^{-\mathrm{i}\pi/3},$$

and hence

$$a_n = A\mathrm{e}^{n\mathrm{i}\pi/3} + B\mathrm{e}^{-n\mathrm{i}\pi/3}.$$

The initial conditions $a_1 = 1$ and $a_2 = 0$ imply that

$$1 = A\mathrm{e}^{\mathrm{i}\pi/3} + B\mathrm{e}^{-\mathrm{i}\pi/3}$$

$$0 = A\mathrm{e}^{2\mathrm{i}\pi/3} + B\mathrm{e}^{-2\mathrm{i}\pi/3},$$

from which it follows that

$$A = \frac{1}{2}\left(1 - \mathrm{i}\frac{1}{\sqrt{3}}\right)$$

$$B = \frac{1}{2}\left(1 + \mathrm{i}\frac{1}{\sqrt{3}}\right).$$

Hence

$$a_n = \tfrac{1}{2}(\mathrm{e}^{n\mathrm{i}\pi/3} + \mathrm{e}^{-n\mathrm{i}\pi/3}) + \frac{\mathrm{i}}{2\sqrt{3}}(-\mathrm{e}^{n\mathrm{i}\pi/3} + \mathrm{e}^{-n\mathrm{i}\pi/3})$$

$$= \cos(n\pi/3) + \frac{1}{\sqrt{3}}\sin(n\pi/3).$$

(This can also be written in the form $a_n = (2/\sqrt{3})\sin[((n+1)/3)\pi]$.

6.3.4. (a) We need to show that the vector equation

$$\alpha_1\{x_1^n\} + \alpha_2\{x_2^n\} + \ldots + \alpha_t\{x_t^n\} = \{0\} \qquad \text{(S6.1)}$$

has just the one solution $\alpha_1 = \alpha_2 = \ldots = \alpha_t = 0$.

If equation (S6.1) holds, then the first t terms of the sequence that occurs on the left-hand side of the equation must be equal to 0. Thus, $\alpha_1, \alpha_2, \ldots, \alpha_t$ must satisfy the system of linear equations,

$$\begin{aligned} \alpha_1 x_1 + \alpha_2 x_2 + \ldots + \alpha_t x_t &= 0 \\ \alpha_1 x_1^2 + \alpha_2 x_2^2 + \ldots + \alpha_t x_t^2 &= 0 \\ \vdots \qquad\qquad & \quad \vdots \\ \alpha_1 x_1^t + \alpha_2 x_2^t + \ldots + \alpha_t x_t^t &= 0. \end{aligned}$$

We can rewrite this in matrix form as

$$\mathbf{A}_t \mathbf{v}_t = \mathbf{0}, \qquad \text{(S6.2)}$$

where

$$\mathbf{A}_t = \begin{pmatrix} x_1 & x_2 & \dots & x_t \\ x_1^2 & x_2^2 & \dots & x_t^2 \\ \vdots & \vdots & & \vdots \\ x_1^t & x_2^t & \dots & x_t^t \end{pmatrix} \quad \text{and} \quad \mathbf{v}_t = \begin{pmatrix} \alpha_1 \\ \alpha_2 \\ \vdots \\ \alpha_t \end{pmatrix}.$$

It is straightforward to show (or a standard algebraic fact that you already know) that

$$\det(\mathbf{A}_t) = x_1 x_2 \dots x_t \prod_{1 \leq i < j \leq t} (x_j - x_i).$$

It follows that if $x_1, x_2, \dots, x_t$ are all different, $\det(\mathbf{A}_t) \neq 0$. Hence $\mathbf{A}_t$ is invertible, and so it follows from equation (S6.2) that

$$\mathbf{v}_t = \mathbf{A}_t^{-1}\mathbf{0} = \mathbf{0}$$

and hence that

$$\alpha_1 = \alpha_2 = \dots = \alpha_t = 0.$$

This completes the proof that the given set of sequences from $\mathbb{R}^\infty$ is linearly independent.

(b) Suppose, first, that

$$\alpha_1\{y_n^1\} + \alpha_2\{y_n^2\} + \dots + \alpha_k\{y_n^k\} = \{0\}.$$

It follows that, for $1 \leq n \leq k$, the nth term of the sequence on the left-hand side of this equation equals 0, that is,

$$\alpha_1 y_n^1 + a_2 y_n^2 + \dots + \alpha_k y_n^k = 0.$$

Since $y_n^i = 0$, for $i \neq n$, and $y_n^n = 1$, it follows that

$$\alpha_n = 0.$$

Thus the given set of sequences is linearly independent.

Now suppose that $\{x_n\}$ is a solution of the recurrence relation. The sequence $\{z_n\}$ defined by

$$\{z_n\} = x_1\{y_n^1\} + x_2\{y_n^2\} + \dots + x_k\{y_n^k\}$$

is a solution of the recurrence relation, by Theorem 6.2, and for $1 \leq n \leq k$, $z_n = x_n$. It follows that $\{z_n\} = \{x_n\}$. Thus the sequence $\{x_n\}$ is a linear combination of the sequences $\{y_n^1\}, \{y_n^2\}, \dots, \{y_n^k\}$, so these sequences span the solution space. Hence, as they form a linearly independent spanning set, they form a basis for the solution space.

(c) It follows from (b) that the solution space of the recurrence relation has dimension k. The sequences $\{x_1^n\}, \dots, \{x_k^n\}$ form a set of k sequences from the solution space and, by (a), they are linearly independent. Therefore they form a basis for the solution space. Hence each solution of the recurrence relation can be written as a linear combination of them.

6.3.5. We show first that each of the given sequences is a solution of the recurrence relation. Let $p(x)$ be the polynomial

$$\alpha_0 x^k + \alpha_1 x^{k-1} + \ldots + \alpha_{k-1}x + \alpha_k$$

We already know that since x_0 is a solution of the equation $p(x) = 0$, then the sequence $\{x_0^n\}$ is a solution of the recurrence relation.

We define the sequence of polynomials $p_0, p_1, \ldots, p_{r-1}$ by

$$p_0(x) = x^{n-k} p(x)$$

and, for $1 \leqslant i < r$,

$$p_i(x) = xp'_{i-1}(x),$$

where the prime denotes differentiation with respect to x.

Since x_0 is a zero of order r of the polynomial p, it is also a zero of order r of p_0. The order of this zero decreases by 1 with each differentiation. So, for $1 \leqslant i < r$, x_0 is a zero of order $r - i$. In particular it follows that x_0 is a zero of each of the polynomials p_i, for $1 \leqslant i < r$. Now it is straightforward to check that, for $1 \leqslant i < r$,

$$p_i(x) = \alpha_0 n^i x^n + \alpha_1 (n-1)^i x^{n-1} + \ldots \alpha_k (n-k)^i x^{n-k}$$

and hence it follows that, for $1 \leqslant i < r$,

$$\alpha_0 n^i x_0^n + \alpha_1 (n-1)^i x_0^{n-1} + \ldots + \alpha_k (n-k)^i x_0^{n-k} = 0.$$

We have thus shown that each of the sequences $\{x_0^n\}$, $(nx_0^n\}$, ..., $\{n^{r-1}x_0^n\}$ are solutions of the recurrence relation.

We complete the solution by showing that these sequences are linearly independent. We do this using the method used in the solution to Exercise 6.3.4(a). The matrix we arrive at in this case, corresponding to the matrix $\mathbf{A}_t$ in the above solution, is

$$\mathbf{B}_r = \begin{bmatrix} x_0 & 1x_0 & 1^2 x_0 & \ldots & 1^{r-1}x_0 \\ x_0^2 & 2x_0^2 & 2^2 x_0^2 & \ldots & 2^{r-1}x_0^2 \\ \vdots & \vdots & \vdots & & \vdots \\ x_0^r & rx_0^r & r^2 x_0^r & \ldots & r^{r-1}x_0^r \end{bmatrix}.$$

We see that

$$\det(B_r) = x_0^{r(r-1)/2} \begin{vmatrix} 1 & 1 & 1 & \dots & 1 \\ 1 & 2 & 2^2 & \dots & 2^{r-1} \\ \vdots & \vdots & \vdots & & \vdots \\ 1 & r & r^2 & \dots & r^{r-1} \end{vmatrix}$$

$$= x_0^{r(r-1)/2} \prod_{1 \leq i < j \leq r} (j - i) \neq 0$$

(Here we have used essentially the same factorization as we used in Exercise 6.3.4(a).) Since $\det(\mathbf{B}_r) \neq 0$, it follows that the sequences $\{x_0^n\}, \{nx_0^n\}, \dots, \{n^{r-1}x_0^n\}$ form a linearly independent set.

6.4.1. (a) A sequence of length $n+1$ containing an even number of vowels has one of the two forms c$_e$ or v$_o$ where c is a consonant and v is a vowel, $_e$ is a sequence of length n containing an even number of vowels, and $_o$ is a sequence of length n containing an odd number of vowels. c can be chosen in 21 ways and v in five ways. $_e$ can be chosen in a_n ways. Hence, as there are altogether 26^n sequences of length n, $_o$ can be chosen in $(26^n - a_n)$ ways. It follows that

$$a_{n+1} = 21a_n + 5(26^n - a_n) = 16a_n + 5(26^n),$$

and, clearly, $a_1 = 21$.

(b) Let A be the generating function for the sequence $\{a_n\}$. It follows from the recurrence relation that

$$\frac{1}{x} \sum_{n=1}^{\infty} a_{n+1}x^n = 16 \sum_{n=1}^{\infty} a_n x^n + 5 \sum_{n=1}^{\infty} 26^n x^n.$$

Thus

$$\frac{1}{x}(A - 21x) = 16A + 5\left(\frac{26x}{1 - 26x}\right).$$

It follows that

$$A = \frac{21x}{1 - 16x} + \frac{130x^2}{(1 - 16x)(1 - 26x)} = \frac{21x - 416x^2}{(1 - 16x)(1 - 26x)}.$$

(c) Using the partial fraction method, we see that

$$A = \frac{1}{2}\left(\frac{1}{1 - 16x} + \frac{1}{1 - 26x}\right) - 1$$

$$= \frac{1}{2}\left(\sum_{n=0}^{\infty} 16^n x^n + \sum_{n=0}^{\infty} 26^n x^n\right) - 1.$$

Hence, equating coefficients, we see that for $n \geq 1$,

$$a_n = \tfrac{1}{2}(16^n + 26^n).$$

6.4.2. We let r_n^m be the number of arithmetical operations needed to row reduce an $m \times n$ matrix to echelon form. Consider such a matrix:

$$\begin{bmatrix} a_{11} & a_{12} & a_{13} & \cdots & a_{1n} \\ a_{21} & a_{22} & a_{23} & \cdots & a_{2n} \\ \vdots & \vdots & \vdots & & \vdots \\ a_{m1} & a_{m2} & a_{m3} & \cdots & a_{mn} \end{bmatrix}$$

The first step is to ensure, by interchanging rows if necessary, that $a_{11} \neq 0$. (If all the entries in the first column are 0, we are really dealing with an $m \times (n-1)$ matrix, but we are aiming to calculate the value of r_n^m in the worst case.) This does not require any arithmetic. We then divide the top row by a_{11} to get 1 in the first row and column. This involves $(n-1)$ divisions. (We do not need to do any arithmetic to work out that $a_{11}/a_{11} = 1$.) Then, for $2 \leqslant s \leqslant m$, we multiply the first row by a_{s1} and subtract the result from the sth row. Without doing any arithmetic we know that this makes the first entry in each of these rows 0. For each of the remaining $(n-1)$ entries this involves doing one multiplication and one subtraction, so, altogether $2(n-1)(m-1)$ arithmetical operations are involved. We will now have carried out in total $(n-1) + 2(n-1)(m-1) = (n-1)(2m-1)$ arithmetical operations and we will have obtained a matrix of the form

$$\begin{bmatrix} 1 & a'_{12} & a'_{13} & \cdots & a'_{1n} \\ 0 & a'_{22} & a'_{23} & \cdots & a'_{2n} \\ \vdots & \vdots & \vdots & & \vdots \\ 0 & a'_{m2} & a'_{m3} & \cdots & a'_{mn} \end{bmatrix}$$

At the second step we ensure that a'_{22} is non-zero, divide the second row by a'_{22} and then subtract suitable multiples of the second row from the remaining rows, so that all the entries in the second column, except for the one in the second row, are 0. This takes the same amount of arithmetic as the first step except that we can now ignore the first column, so that we are essentially dealing with an $m \times (n-1)$ matrix. Hence this requires $(n-2)(2m-1)$ steps.

Assuming the worst possible case where the matrix has rank m, we have to carry out m steps in this process and so

$$r_n^m = (n-1)(2m-1) + (n-2)(2m-1) + \ldots + (n-m)(2m-1) = \tfrac{1}{2}m(2n-m-1)(2m-1).$$

6.4.3. Since the indicated method for calculating the inverse of an $n \times n$ matrix involves row-reducing an $n \times 2n$ matrix to echelon form, it is tempting simply to use the result of Exercise 6.4.2, and say that the number of arithmetic operations required is

$$r_{2n}^n = \tfrac{1}{2}n(3n-1)(2n-1).$$

However, this does not take into account the special form of the matrix $(\mathbf{AI}_n)$ that we row-reduce. If you take this into account you will see that each step involves the same amount of arithmetic, names n divisions, followed by $(n-1)n$ multiplications and then $(n-1)(n-1)$ subtractions, making $(2n^2-2n+1)$ operations in all. Since n steps have to be carried out, the total number of arithmetical operations involved is $n(2n^2-2n+1)$. (Note that this is asymptotic to $2n^3$, whereas r_{2n}^n is asymptotic to $3n^3$.)

6.4.4. *The Towers of Hanoi*

When there is just one disc only one move is needed to transfer it to the third peg, so $a_1 = 1$. To move $(n+1)$ discs to the third peg we need first to transfer the top n discs to the second peg. This takes a_n moves. Then in one move we can transfer the bottom, largest disc, to the third peg. Finally we have to transfer the n other discs from the second to the third peg. This takes another a_n moves. Hence, the required recurrence relation is

$$a_{n+1} = 2a_n + 1. \tag{S6.3}$$

This is a particularly simple non-homogeneous linear recurrence relation. The general solution of the associated homogeneous equation $a_{n+1} = 2a_n$, is $a_n = A2^n$. If we try a particular solution of equation (S6.3) of the form $a_n = c$, we see that the constant c has to satisfy

$$c = 2c + 1$$

and hence $c = -1$. So the general solution of equation (S6.3) is $a_n = A2^n - 1$, and since $a_1 = 1$, it follows that $A = 1$. Hence

$$a_n = 2^n - 1.$$

It follows that with $n = 64$, the process takes $2^{64} - 1$ steps. $2^{64} - 1$ seconds comes to about 5.8×10^{11} years.

6.5.1. Let $A(x) = \Sigma_{n=1}^{\infty} a_n x^n$. From the recurrence relation

$$a_n = \sum_{k=1}^{n-1} a_k a_{n-k} \qquad \text{for } n \geqslant 2$$

it follows that

$$\sum_{n=2}^{\infty} a_n x^n = \sum_{n=2}^{\infty} \left(\sum_{k=1}^{n-1} a_k a_{n-k} \right) x^n.$$

That is,

$$A(x) - x = (A(x))^2$$

and hence

$$(A(x))^2 - A(x) + x = 0,$$

from which it follows, from the formula for the solution of a quadratic equation, that

$$A(x) = \frac{1 \pm \sqrt{1 - 4x}}{2}.$$

Since $A(x)$ has no constant term, we need to take the + sign here, and so

$$\begin{aligned} A(x) &= \tfrac{1}{2}(1 - \sqrt{1 - 4x}) \\ &= \frac{1}{2}\left(1 - \left(1 - \frac{1}{2}4x - \sum_{n=2}^{\infty} \frac{(2n-3)!(4x)^n}{n!(n-2)!2^{2n-2}}\right)\right) \end{aligned}$$

using the power series for $\sqrt{1-x}$ as given in section 6.5. Therefore

$$\begin{aligned} A(x) &= x + \frac{1}{2} \sum_{n=2}^{\infty} \frac{(2n-3)!(4x)^n}{n!(n-2)!2^{2n-2}} \\ &= x + \sum_{n=2}^{\infty} \frac{2(2n-3)!x^n}{n!(n-2)!} \end{aligned}$$

It follows that for $n \geqslant 2$,

$$a_n = \frac{2(2n-3)!}{n!(n-2)!}.$$

Thus, using Sterling's approximation for $n!$,

$$
\begin{aligned}
a_n &= \frac{2(2n-3)!}{n!(n-2)!} \\
&= \frac{2n(n-1)(2n)!}{2n(2n-1)(2n-2)(n!)^2} \\
&\sim \frac{(n-1)\left(\frac{2n}{\mathrm{e}}\right)^{2n}\sqrt{4\pi n}}{(2n-1)(2n-2)\left(\left(\frac{n}{\mathrm{e}}\right)^{n}\sqrt{2\pi n}\right)^2} \\
&\sim \frac{2^{2n}}{4\sqrt{\pi} n^{3/2}}.
\end{aligned}
$$

Thus the constant α has the value $1/(4\sqrt{\pi})$.

6.6.1.
$$\frac{1}{10}\left(\frac{1}{x-7} - \frac{1}{x+3}\right).$$

6.6.2.
$$\frac{1}{9}\left(\frac{2}{x+1} - \frac{6}{(x+1)^2} + \frac{7}{x-5}\right).$$

6.6.3.
$$1 + \frac{2}{x^2+1} - \frac{5}{x^2+2}.$$

6.6.4.
$$\frac{1}{2}\left(\frac{1}{(x+1)^2} + \frac{1}{(x+1)} - \frac{x}{(x^2+1)}\right).$$

CHAPTER 7

7.1.1. (a) (1 3 8 2)(4 9 6)(5 7 10).
(b) (1 5 7 4 10)(3 8 9).

7.1.2. The cycle of length 4 can be written in four different ways, and each of the cycles of length 3 in three different ways. Also the three different cycles can be ordered in $3! = 6$ ways. Hence the total number of different ways of writing the permutation is $4 \times 3 \times 3 \times 6 = 216$.

7.1.3. We know that S_n contains $n!$ different permutations. How many of these consist of a single cycle of length n? We can arrange the n numbers from 1 to n in a bracket in $n!$ different ways. But each cycle of length n can be written in n different ways. Hence there are $n!/n = (n-1)!$ different cycles of length n. So the probability that an element from S_n consists of a single cycle of length n is $(n-1!)/n! = 1/n$.

7.2.1. (a) (1 7 3)(2 9)(5 6).
(b) (1 2 7 5)(4 9 8 6).

7.2.2. (a) Suppose both e_1 and e_2 satisfy the property of being an identity element. Then we have

$$e_1 = e_1 e_2 = e_2.$$

(b) Consider the product y^*x^*z, evaluated in two different ways. First, we have

$$(y^*x)^*z = e^*z = z,$$

and second,

$$y^*(x^*z) = y^*e = y.$$

By the associativity property, $(y^*x)^*z = y^*(x^*z)$. Thus $y = z$.

7.2.3. (a) Suppose $x, y, z \in G$ are such that $x^*y = x^*z$. Since G is a group, x has an inverse x^{-1} in G. Since $x^*y = x^*z$ it follows that $x^{-1*}(x^*y) = x^{-1*}(x^*z)$, and hence $(x^{-1*}x)^*y = (x^{-1*}x)^*z$, that is, $y = z$.
(b) The proof of this is exactly similar to that of part (a).

7.2.4. First, we note that, in general, permutation groups are not commutative. Thus, for example, (1 2 3)(1 2) = (1 3), whereas (1 2)(1 2 3) = (2 3). This shows that the group S_n, for $n \geqslant 3$, is not commutative (as we can regard (1 2 3) and (1 2) as permutations from S_n provided that $n \geqslant 3$). Exceptionally, S_1 and S_2 are commutative.

The groups $(\mathbb{Z}, +)$, $(\mathbb{Q}, +)$, $(\mathbb{R}, +)$, $(\mathbb{C}, +)$, $(\mathbb{Q}^*, \times)$, $(\mathbb{R}^*, \times)$ and $(\mathbb{C}^*, \times)$ are all commutative.

In general, matrix multiplication is not commutative. So for $n \geqslant 2$, the group M_n of invertible $n \times n$ matrices is not commutative.

The groups $(\mathbb{Z}_n, +)$ are all commutative.

The abstract group whose Cayley table is given in section 7.2 can be seen to be non-commutative, for example, $ar \neq ra$.

7.2.5.

	e	a	b	c	d
e	e	a	b	c	d
a	a	e	c	d	b
b	b	d	e	a	c
c	c	b	d	e	a
d	d	c	a	b	e

It is easily seen that this table satisfies the closure condition, e

acts as an identity element, and each element acts as its own inverse. However, this is *not* the Cayley table of a group as it does not satisfy the associativity condition. For example,

$$(ab)d = cd = a$$

whereas

$$a(bd) = ac = d.$$

7.3.1. First note than an isometry, f, must be injective. For if $p \neq q$, then $d(p, q) > 0$, and hence $d(f(p), f(q)) > 0$, and so $f(p) \neq f(q)$. Since a symmetry is surjective, it follows that a symmetry, f, of a figure $\mathcal{F}$ is a bijection, $f: \mathcal{F} \to \mathcal{F}$. Hence f has an inverse $f^{-1}: \mathcal{F} \to \mathcal{F}$, which is also a bijection. f^{-1} is also an isometry, since, using the fact that f is an isometry,

$$d(f^{-1}(p), f^{-1}(q)) = d(f(f^{-1}(p)), f(f^{-1}(q))) = d(p, q).$$

Hence f^{-1} is also a symmetry of $\mathcal{F}$.

7.3.2.

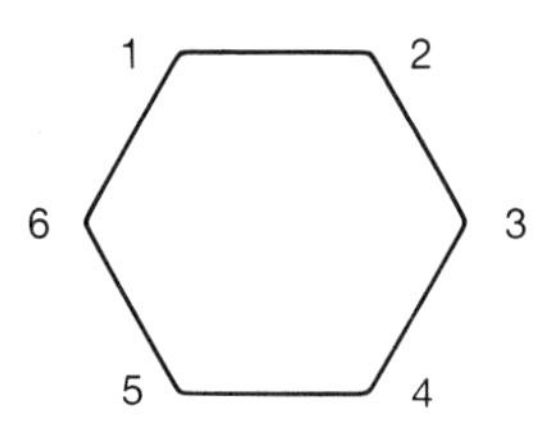

7.3.2. A regular hexagon has 12 symmetries. In addition to the identity, e, there are five rotations, r_1, r_2, r_3, r_4 and r_5, where r_t is a rotation through t-sixths of a complete turn clockwise, i.e. through an angle $2\pi t/6$ clockwise. There are three reflections, q_1, q_2, q_3 in axes joining opposite vertices, and three reflections p_1, p_2, p_3 in axes joining the midpoints of opposite edges. With the original positions of the vertices of the hexagon labelled as shown, the permutations corresponding to these symmetries are as follows:

e e
r_1 $(1\,2\,3\,4\,5\,6)$
r_2 $(1\,3\,5)(2\,4\,6)$
r_3 $(1\,4)(2\,5)(3\,6)$
r_4 $(1\,5\,3)(2\,6\,4)$
r_5 $(1\,6\,5\,4\,3\,2)$
q_1 $(2\,6)(3\,5)$

q_2 (1 3)(4 6)
q_3 (1 5)(2 4)
p_1 (1 2)(3 6)(4 5)
p_2 (1 4)(2 3)(5 6)
p_3 (1 6)(2 5)(3 4)

In general, a regular n-sided figure has $2n$ symmetries, n rotations (including the identity) and n reflections. The rotations are through $2\pi t/n$ for $t = 0, 1, 2, \ldots, n-1$. If n is even, half the reflections are in axes joining opposite vertices, and half in axes joining the midpoints of opposite edges. If n is odd all the reflections are in axes joining a vertex to the midpoint of the opposite edge. [The group of these $2n$ symmetries is sometimes called the **dihedral** group of order $2n$, and the notation D_n is used for it.]

7.3.3. It will help to answer this question, and to understand this solution, if you find, or make for yourself, a regular tetrahedron and a cube.

(a) First of all we have the identity, as always.

There are two different kinds of axis of rotational symmetry. The first consists of axes joining a vertex to the midpoint of an opposite face. There are four of these axes, and two rotational symmetries about each of these axes, through an angle $2\pi/3$ clockwise and through $2\pi/3$ anti-clockwise, making eight of these rotations altogether. There are also three axes of rotational symmetry joining the midpoints of opposite edges, with one rotation through π about each of these axes (a rotation through π clockwise is the same as a rotation, about the same axis, anti-clockwise). This makes 11 rotational symmetries in all.

There are six reflectional symmetries in planes through two vertices and the midpoint of the opposite edge.

The remaining six symmetries are more difficult to find, as they consist of a rotation combined with a reflection. If we use the numbers 1, 2, 3, 4 to label the positions of the vertices, these symmetries correspond to cycles of length 4.

Thus, altogether the regular tetrahedron has 24 symmetries. This is the same as the number of permutations in S_4. It can thus be seen that each permutation in S_4 corresponds to a symmetry of the tetrahedron. (*Warning*: this is a very special case—in general a figure with n vertices has fewer than $n!$ symmetries. For example, we have already seen that a square has only eight symmetries.)

(b) We classify these as follows:

Type of rotation	*Number of this type*
The identity.	1
Rotations through $\pi/2$, both clockwise and anti-clockwise, about axes joining the midpoints of opposite faces.	6
Rotations through π, about axes joining the midpoints of opposite faces.	3
Rotations through π, about axes joining the midpoints of opposite edges.	6
Rotations through $2\pi/3$, both clockwise and anti-clockwise, about axes joining opposite vertices.	8

This makes 24 rotational symmetries in all. (The cube has a further 24 symmetries which are either reflections, or combinations of reflections and rotations.)

7.4.1. First, the easy part. If H is a subgroup of G and $g, h \in H$, then, since H satisfies the inverse condition $h^{-1} \in H$, and hence, as H satisfies the closure condition $gh^{-1} \in H$.

For the converse, suppose that H is non-empty and that,

$$\text{for all } g,\ h \in H,\ gh^{-1} \in H. \qquad (*)$$

Since $H \neq \varnothing$, there is at least one element $h_0 \in H$. Hence, by the condition $(*)$, with $g = h = h_0$, we have $h_0 h_0^{-1} = e \in H$. Thus H contains the identity element of G. It follows, for each $h \in H$, using condition $(*)$ with $g = e$, that $h^{-1} \in H$. Hence H also satisfies the inverse condition. Finally suppose that $g, h \in H$. Then by what we have just proved, $h^{-1} \in H$. Hence, from the condition $(*)$, with h replaced by h^{-1}, $g(h^{-1})^{-1} \in H$, that is $gh \in H$. Hence H also satisfies the closure condition. This completes the proof that H is a subgroup of G.

7.4.2. It is straightforward to check, from the Cayley table for G, that $\{e, b\}$ is a subgroup of G. Its distinct cosets are $\{e, b\}$, $\{a, c\}$, $\{p, q\}$, $\{r, s\}$.

7.4.3. S_4 is a group of order 24. Therefore the possible orders of its subgroups are 1, 2, 3, 4, 6, 8, 12 and 24. We list all the subgroups of G together with just a few hints about finding them. The more you know about group theory the easier it is to tackle a problem of this kind. Do not worry if you do not understand all the explanations. We will be using some of the theory of section 7.5 and some group theory which is not explained in this book.

Subgroups of order 1: Just the trivial subgroup $\{e\}$.
Subgroups of order 2: A group of order 2 consists of the identity, and an element of order 2. So we get all the subgroups of order 2 by taking the identity, e, together with each element of S_4 of order 2 in turn. So the nine subgroups of order 2 are:

$$\{e, (12)\}, \{e, (13)\}, \{e, (14)\}, \{e, (23)\}, \{e, (24)\},$$

$$\{e, (34)\}, \{e, (12)(34)\}, \{e, (13)(24)\}, \{e, (14)(23)\}.$$

Subgroups of order 3: A group of order 3 consists of the identity, and an element of order 3 together with its inverse. So the four subgroups of order 3 are:

$$\{e, (123), (132)\}, \{e, (124), (142)\},$$

$$\{e, (134), (143)\}, \{e, (234), (243)\}.$$

Subgroups of order 4: A group of order 4 is either a cyclic group, generated by an element of order 4, or consists of the identity and three elements of order 2, each of which is obtained by combining the other two. S_4 has three cyclic subgroups of order 4:

$$\{e, (1234), (13)(24), (1432)\}, \{e, (1324), (12)(34), (1423)\},$$

$$\{e, (1342), (14)(23), (1243)\}.$$

S_4 also has four subgroups of order 4 consisting of the identity and three elements of order 2:

$$\{e, (12)(34), (13)(24), (14)(23)\}, \{e, (12), (34), (12)(34)\},$$

$$\{e, (13), (24), (13)(24)\}, \{e, (14), (23), (14)(23)\}.$$

Subgroups of order 6: A group of order 6 is either cyclic, or is isomorphic to S_3. There are no cyclic subgroups of order 6, as S_4 does not contain any elements of order 6. There are four subgroups of S_4 which are isomorphic to S_3:

$$\{e, (12), (13), (23), (123), (132)\},$$

$$\{e, (12), (14), (24), (124), (142)\},$$

$$\{e, (13), (14), (34), (134), (143)\},$$

$$\{e, (23), (24), (34), (234), (243)\}.$$

Subgroups of order 8: Life becomes more difficult when it comes to groups of order 8, since, in general there are many possibilities for groups of this order (technically speaking, there are five non-isomorphic groups of order 8). It is helpful to tackle the problem of finding the subgroups of S_4 of order 8 by using the theory of section 7.5.

By Theorem 7.14 a subgroup of S_4 of order 8 cannot contain any elements of order 3, and so, apart from the identity, must be made up of elements of orders 2 and 4. If any two of the elements (12), (13), (14) are combined, the result is an element of order 3. So a subgroup of order 8 can contain only one of these elements, and likewise only one of (21), (23), (24), etc. It follows that such a subgroup cannot consist of the identity and seven elements of order 2. Similarly if a cycle of length 4, such as (1234), is combined with any other such cycle, other than its inverse, the result is an element of order 3.

It follows from these considerations, that a subgroup of S_4 of order 8 must consist of the identity, an element of order 4 and its inverse, and five elements of order 2. Thus of the five possibilities for groups of order 8, thc only possibility for a subgroup of S_4 of order 8 is a subgroup which is isomorphic to the symmetry group of a square. S_4 has three such subgroups:

$$\{e, (1234), (13)(24), (1432), (13), (24), (12)(34), (14)(23)\},$$

$$\{e, (1324), (12)(34), (1423), (12), (34), (13)(24), (14)(23)\},$$

$$\{e, (1243), (14)(23), (1342), (14), (23), (12)(34), (13)(24)\}.$$

Subgroups of order 12: Here we use some group theory not mentioned elsewhere in this book. A subgroup of 12 elements from a group of order 24 must be a normal subgroup, and hence must be a union of conjugacy classes. Hence if it contains one element of a particular cycle type, it must contain all the elements of that cycle type. It follows that S_4 has just one subgroup of order 12:

$$\{e, (123), (132), (124), (142), (134), (143), (234),$$
$$(243), (12)(34), (13)(24), (14)(23)\}.$$

Subgroups of order 24: The whole group S_4 itself.

7.4.4. The given subgroup, H, is a subgroup of S_4, so any subgroup of H of order 6 would be a subgroup of S_4 of order 6. From the answer to Exercise 7.4.3, above, it can be seen that none of the subgroups of S_4 of order 6 is a subgroup of H. Hence H has no subgroups of order 6.

7.5.1. The identity has order 1. The elements b, p, q, r, s have order 2, and a and c have order 4.

7.5.2.

Elements	*Order*
e	1
(12), (13), (14), (23), (24), (34)	2
(12)(34), (13)(24), (14)(23)	2
(123), (132), (124), (142), (134), (143), (234), (243)	3
(1234), (1432), (1324), (1423), (1243), (1342)	4

7.5.3. Suppose G is a group in which every element, other than the identity, is of order 2. Take $g, h \in G$. Then gh is an element of G, and hence $(gh)^2 = e$. Thus $(gh)^{-1} = gh$. Now, in any group, $(gh)^{-1} = h^{-1}g^{-1}$, and as the elements of G have order 2, $h^{-1} = h$ and $g^{-1} = g$. Thus $hg = gh$. This proves that G is commutative.

7.6.1. It is straightforward to check that the permutation in S_{52} corresponding to the over riffle shuffle has cycle type $x_1^2x_2x_8^6$, and hence has order $= \text{lcm}\{1, 2, 8\} = 8$. The permutation corresponding to an under riffle shuffle has cycle type x_{52}, and hence has order 52.

7.6.2.

n	*Cycle type of element of* S_n *of highest order*	*Order of this element*
1	x_1	1
2	x_2	2
3	x_3	3
4	x_4	4
5	x_2x_3	6
6	$x_1x_2x_3$ or x_6	6
7	x_3x_4	12
8	x_3x_5	15
9	x_4x_5	20
10	$x_2x_3x_5$	30

7.6.3. In trying to find a permutation in S_n of highest order, we are seeking a partition of n, say,

$$k_1 + k_2 + \ldots + k_t = n$$

for which $\text{lcm}\{k_1, \ldots, k_t\}$ is as large as possible. It is easy to see that to achieve this we need only consider partitions where the numbers $k_1, \ldots, k_t$ are co-prime. Hence the only numbers we need consider are prime numbers, powers of prime numbers, as well as the number 1. This cuts down the amount of searching. It then turns out that the highest order of an element of S_{52} is 180 180, corresponding to an element of S_{52} of cycle type

$$x_1^3 x_4 x_5 x_7 x_9 x_{11} x_{13},$$

which itself corresponds to the partition

$$13 + 11 + 9 + 7 + 5 + 4 + 1 + 1 + 1 = 52.$$

CHAPTER 8

8.1.1. Since an 8×8 chessboard has 64 squares, there are 2^{64} different colourings using two colours. $2^{64} = 1.8 \times 10^{19}$.

8.1.2. Consider the following two colourings:

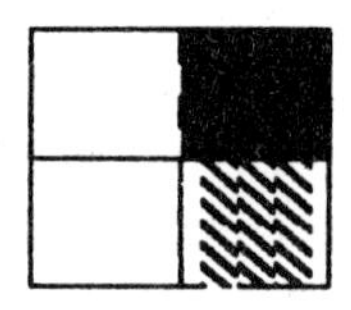

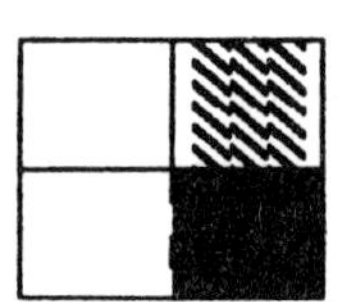

Each colouring can be obtained from the other by a reflection in the horizontal axis of symmetry, but there is no way of rotating one colouring so that it looks like the other one.

8.2.1. (a) The identity element of the group $(\mathbb{R}^*, \times)$ is the number 1. $1 \cdot (x, y) = (x, y)$ and hence GA1 is satisfied. Now suppose, $a, b \in \mathbb{R}$, and $(x, y) \in \mathbb{R}^2$. Then

$$(ab) \cdot (x, y) = (abx, aby) = (a(bx), a(by)) = a \cdot (bx, by)$$
$$= a \cdot (b \cdot (x, y)).$$

Thus GA2 also holds. So the group action axioms both hold in this case.

(b) The identity element of the group $(\mathbb{R}, +)$ is the number 0. So the first axiom GA1 does not hold, as unless $(x, y) = (0, 0)$, $0 \cdot (x, y) = (0, 0) \neq (x, y)$.

(c) There is no need for any elaborate algebra in this case, as it can be seen that the group action is just the usual operation of matrix multiplication, if we write the elements of $\mathbb{R}^2$ as column matrices. Thus GA1 just boils down to the fact that if $\mathbf{I}$ is the 2×2 identity matrix, then

$$\mathbf{I}\begin{pmatrix} x \\ y \end{pmatrix} = \begin{pmatrix} x \\ y \end{pmatrix}, \text{ for all } \begin{pmatrix} x \\ y \end{pmatrix} \in \mathbb{R}^2,$$

and GA2 just amounts to the fact that matrix multiplication is associative. Thus for all $\mathbf{A}, \mathbf{B} \in G$ and $\mathbf{x} \in \mathbb{R}^2$, we have

$$(\mathbf{AB})\mathbf{x} = \mathbf{A}(\mathbf{Bx}),$$

which we can rewrite, in terms of the group action, as

$$(\mathbf{AB})\cdot\mathbf{x} = \mathbf{A}\cdot(\mathbf{B}\cdot\mathbf{x}).$$

So both of the axioms for a group action hold in this case.

8.2.2. (a) Suppose $x, y \in X$ and $g\cdot x = g\cdot y$. Then $g^{-1}\cdot(g\cdot x) = g^{-1}\cdot(g\cdot y)$, that is, using GA2, $(g^{-1}g)\cdot x = (g^{-1}g)\cdot y$, thus $e\cdot x = e\cdot y$. Hence, using GA1, $x = y$. Thus we have shown that $\phi_g(x) = \phi_g(y)$ implies $x = y$. This proves that ϕ_g is injective (one-to-one).

Next, suppose $y \in X$. Then $g^{-1}\cdot y \in X$, and $g\cdot(g^{-1}\cdot y) = y$. Thus with $x = g^{-1}\cdot y$, $\phi_g(x) = y$. Thus ϕ_g is surjective (onto). This completes the proof that ϕ_g is a permutation of X.

(b) To show that Φ is a homomorphism we need to show that for all $g, h \in G$, $\Phi(gh) = \Phi(g) \circ \Phi(h)$, that is, that $\phi_{gh} = \phi_g \circ \phi_h$. Now for each $x \in X$,

$$\phi_{gh}(x) = gh\cdot x = g\cdot(h\cdot x) = \phi_g(\phi_h(x)) = (\phi_g \circ \phi_h)(x),$$

and this shows that $\phi_{gh} = \phi_g \circ \phi_h$, as required.

8.3.1. We saw that (a) and (c) are examples of group actions.

(a) The orbit of a point $(x, y) \in \mathbb{R}^2$ is given by,

$$\begin{aligned} O_{(x,y)} &= \{g\cdot(x, y) : g \in \mathbb{R}^*\} \\ &= \{(gx, gy) : g \in \mathbb{R}^*\}. \end{aligned}$$

Thus if $(x, y) = (0, 0)$, the orbit just consists of the origin by itself, and otherwise the orbit of (x, y) consists of the line which goes through (x, y) and $(0, 0)$ but with the origin deleted.

(c) In this case it is not so easy to see what the orbits are. So it is best to start by calculating the orbits of particular points.

$$O_{(0,0)} = \{\mathbf{A0} : \mathbf{A} \in G\} = \{\mathbf{0}\},$$

since $\mathbf{A0} = \mathbf{0}$ for every matrix $\mathbf{A}$. So the orbit of the origin just consists of the origin by itself.

$$O_{(1,0)} = \left\{\begin{pmatrix} a & 0 \\ c & b \end{pmatrix}\cdot(1, 0) : a, b, c \in \mathbb{R}, \text{ and } ab \neq 0\right\}$$

$$= \{(a, c) : a, c \in \mathbb{R} \text{ and } a \neq 0\},$$

and so the orbit of $(1, 0)$ consists of the whole of $\mathbb{R}^2$ other than the y-axis.

Since the orbits are equivalence classes this is the orbit of every point not on the y-axis. So to find a new orbit, we need to consider the orbit of a point on the y-axis

$$O_{(0,1)} = \left\{\begin{pmatrix} a & 0 \\ c & b \end{pmatrix}\cdot(0, 1) : a, b, c \in \mathbb{R} \text{ and } ab \neq 0\right\},$$

$$= \{(0, b) : b \in \mathbb{R} \text{ and } b \neq 0\},$$

and so the orbit of $(0, 1)$ consists of the entire y-axis other than the origin.

Since the three orbits we have calculated between them include all the points of $\mathbb{R}^2$, these are the only orbits of this particular action.

8.4.1. (a) Suppose that $g, h \in Z(G)$, and $x \in G$. Using the associativity property, and the fact that g, h commute with x, we have

$$(gh)x = g(hx) = g(xh) = (gx)h = (xg)h = x(gh),$$

which shows that gh commutes with x. So $gh \in Z(G)$. The identity element of G commutes with each element of G, as $ex = xe = x$, for all $x \in G$. Hence $e \in Z(G)$. Finally, if $g \in Z(G)$, then for each $x \in G$, $gx = xg$. Hence $g^{-1}gxg^{-1} = g^{-1}xgg^{-1}$, that is, $xg^{-1} = g^{-1}x$. Thus g^{-1} also commutes with x. So $g^{-1} \in Z(G)$. This completes the proof that $Z(G)$ is a subgroup of G.

(b)

$$\begin{aligned} g \in Z(G) &\Leftrightarrow \text{for all } x \in G,\ gx = xg \\ &\Leftrightarrow \text{for all } x \in G,\ xgx^{-1} = g \\ &\Leftrightarrow g \text{ is conjugate to itself alone.} \end{aligned}$$

(c) We have seen that the number of elements in a conjugacy class is a divisor of the number of elements in the group. So if $\#(G) = p^k$, then the number of elements in a conjugacy class is either 1 or some power of p. Let there be s conjugacy classes containing just one element. Since the conjugacy classes partition G, $\#(G)$ is the sum of the numbers of elements in the conjugacy classes. The sum of the numbers of elements in the conjugacy classes containing more than one element is a sum of powers of p and hence is a multiple of p, say pm. Thus we have

$$p^k = s + pm.$$

It follows that s is divisible by p, and so $s \geqslant p$. By (b) $s = \#(Z(G))$. Thus $\#(Z(G)) > 1$.

8.4.2. (a) To obtain a p-tuple $(g_0, g_1, \ldots, g_{p-1})$ in X, we can choose the first $p-2$ elements $g_0, \ldots, g_{p-2}$ in any way we like, and the last element g_{p-1} must be $(g_0 g_1 \ldots g_{p-2})^{-1}$. So we have n choices for each of the first $p-2$ elements, and just one choice for the last element. Hence $\#(X) = n^{p-2}$.

(b) It is straightforward to check that for all $j, k \in \mathbb{Z}_p$, and all $(g_0, g_1, \ldots, g_{p-1}) \in X$,

$$j\cdot(k\cdot(g_0, g_1, \ldots, g_{p-1})) = (j + k)\cdot(g_0, g_1, \ldots, g_{p-1}) \qquad \text{(S8.1)}$$

where the addition is modulo p.

Suppose now that $(g_0, g_1, \ldots, g_{p-1}) \in X$, so that $g_0 g_1 \ldots g_{p-1} = e$. Now $1\cdot(g_0, g_1, \ldots, g_{p-1}) = (g_1, g_2, \ldots, g_{p-1}, g_0)$, and

$$g_1 g_2 \ldots g_{p-1} g_0 = g_0^{-1}(g_0 g_1 \ldots g_{p-1}) g_0 = g_0^{-1} e g_0 = e,$$

and hence $1\cdot(g_0, g_1, \ldots, g_{p-1}) \in X$. Thus X is closed under the action of the element 1 of $\mathbb{Z}_p$. If $k \in \mathbb{Z}_p$, $k = 1 + 1 + \ldots + 1$, with k terms in the sum, hence, by equation (S8.1),

$$k\cdot(g_0, g_1, \ldots, g_{p-1}) = 1\cdot(1\cdot \ldots (1\cdot(g_0, g_1, \ldots, g_{p-1})) \ldots)$$

and hence, since X is closed under the action of $1 \in \mathbb{Z}_p$, if $(g_0, g_1, \ldots, g_{p-1}) \in X$, then $k\cdot(g_0, g_1, \ldots, g_{p-1})$ is also in X.

(c) The identity element of $\mathbb{Z}_p$ is 0, from which it is clear that GA1 holds. By equation (S8.1), GA2 holds. Thus we do have a group action.

(d) Clearly, elements of X of the form $(g_0, g_0, \ldots, g_0)$ are fixed by every element of $\mathbb{Z}_p$. If $(g_0, g_1, \ldots, g_{p-1})$ is fixed by the element 1 of $\mathbb{Z}_p$,

$$1\cdot(g_0, g_1, \ldots, g_{p-1}) = (g_0, g_1, \ldots, g_{p-1}).$$

That is,

$$(g_1, g_2, \ldots, g_{p-1}, g_0) = (g_0, g_1, \ldots, g_{p-1}),$$

from which it follows that $g_0 = g_1 = \ldots = g_{p-1}$.

(e) Since the number of elements in each orbit is a divisor of $\#(\mathbb{Z}_p)$, each orbit contains either one element or p elements. Suppose there are s orbits containing just one element, and t orbits containing p elements. Then, as the orbits partition X, $\#(X) = s + pt$, and so, by (a) above,

$$n^{p-2} = s + pt.$$

Hence, if n is divisible by p, s must also be divisible by p. Therefore $s > 1$.

CHAPTER 9

9.1.1. (a) Graphs with one vertex: When there is just one vertex there are no possible edges. Hence there is just one graph with one vertex.

(b) Graphs with two vertices: Here there are two possibilities. The two vertices are either joined or not. So we get two different graphs.

(c) Graphs with three vertices: Here there are three possible edges, and we get four different graphs according as the number of edges is 0, 1, 2 or 3.

(d) Graphs with four vertices: With four vertices there are six possible edges, and in this case it is possible to have graphs with the same number of edges but which are not isomorphic. So altogether there are 11 non-isomorphic graphs with four vertices.

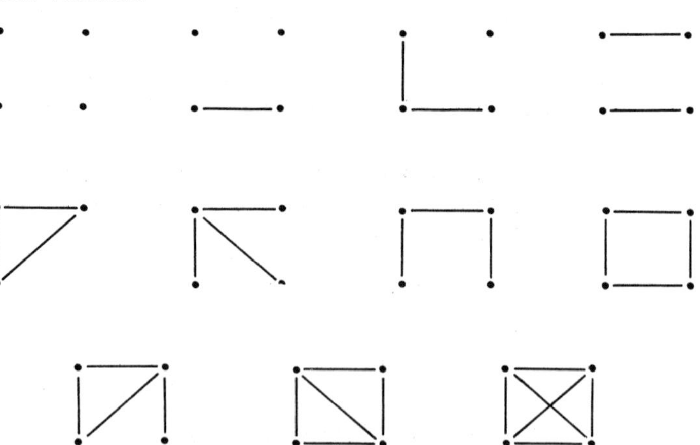

9.2.1. We list the graphs, together with the number of elements in their automorphism groups. The number of distinct labellings then follows from Theorem 9.3.

(a) The graph with one vertex: Clearly this has just one labelling.

(b) Graphs with two vertices: Each of the two graphs with two vertices has two automorphisms, and hence each has just one distinct labelling.

(c) Graphs with three vertices:

Graph $\mathcal{G}$	$\#(\mathcal{A}(\mathcal{G}))$	*Number of distinct labellings*
	6	1
	2	3
	2	3
	6	1

(d) Graphs with four vertices:

Graph $\mathcal{G}$	$\#(\mathcal{A}(\mathcal{G}))$	*Number of distinct labellings*
	24	1
	4	6
	2	12
	8	3
	6	4

(*Continued*)

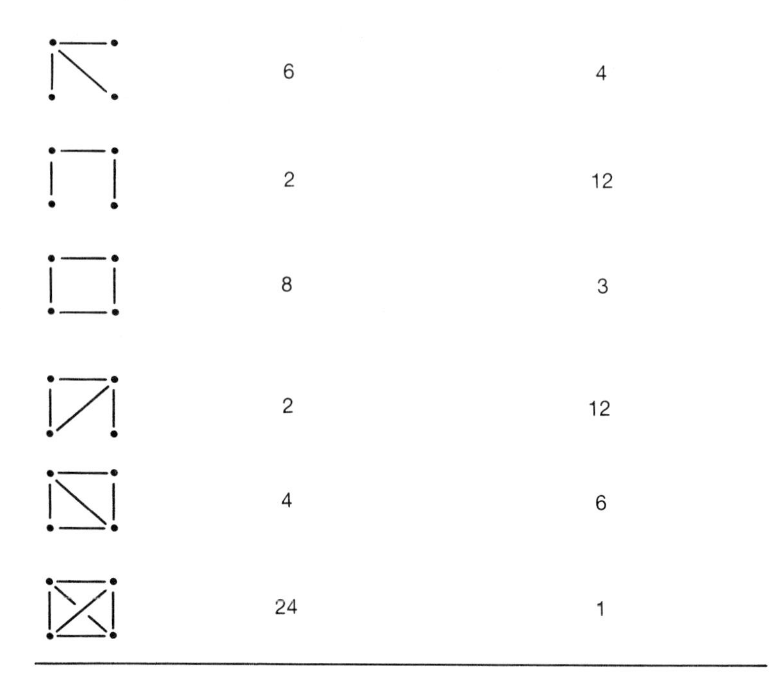

9.2.2. Let $\mathcal{G} = (V, E)$ be a graph. We show first that $\mathcal{A}(\mathcal{G})$ satisfies the closure condition. Suppose $\sigma, \tau \in \mathcal{A}(\mathcal{G})$. Hence they are both bijections from V to V and hence so also is $\sigma\tau$. Also, for all $x, y \in V$,

$$\begin{aligned} xEy &\Leftrightarrow \tau(x)E\tau(y), \text{ as } \tau \text{ is an automorphism of } \mathcal{G}, \\ &\Leftrightarrow \sigma(\tau(x))E\sigma(\tau(y)), \text{ as } \sigma \text{ is an automorphism of } \mathcal{G}, \\ &\Leftrightarrow \sigma\tau(x)E\sigma\tau(y), \text{ by the definition of } \sigma\tau. \end{aligned}$$

Hence $\sigma\tau$ is also an automorphism of $\mathcal{G}$. So $\sigma\tau \in \mathcal{A}(\mathcal{G})$.

It is clear that the identity map $\iota : V \to V$ is an automorphism of $\mathcal{G}$.

Finally, suppose that $\sigma \in \mathcal{A}(\mathcal{G})$. Since σ is a bijection, it has an inverse σ^{-1}, and for all $x, y \in V$,

$$\begin{aligned} xEy &\Leftrightarrow \sigma\sigma^{-1}(x)E\sigma\sigma^{-1}(y) \\ &\Leftrightarrow \sigma^{-1}(x)E\sigma^{-1}(y), \end{aligned}$$

as σ is an automorphism of $\mathcal{G}$. This equivalence shows that σ^{-1} is also an automorphism of $\mathcal{G}$ and hence is in $\mathcal{A}(\mathcal{G})$. So $\mathcal{A}(\mathcal{G})$ contains the inverse of each of its elements.

CHAPTER 10

10.2.1. The cycle type of the identity, e, is x_1^3, and hence $\#(\mathrm{Fix}(e)) = 4^3$. The two rotations of the triangle, p, q, have cycle type x_3, and hence $\#(\mathrm{Fix}(p)) = \#(\mathrm{Fix}(q)) = 4$. Hence, by Burnside's Theorem, the number of different colourings is

$$\tfrac{1}{3}(4^3 + 4 + 4) = 24.$$

10.2.2. In this case we consider the rotations and reflections of a regular octagon. There are eight rotational symmetries: a rotation r_k through $2\pi k/8$, for $0 \leqslant k \leqslant 7$. There are four reflections s_1, s_2, s_3, s_4, in axes joining opposite vertices, and four reflections t_1, t_2, t_3, t_4, in axes joining the midpoints of opposite edges.

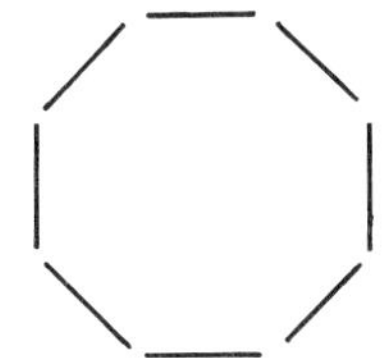

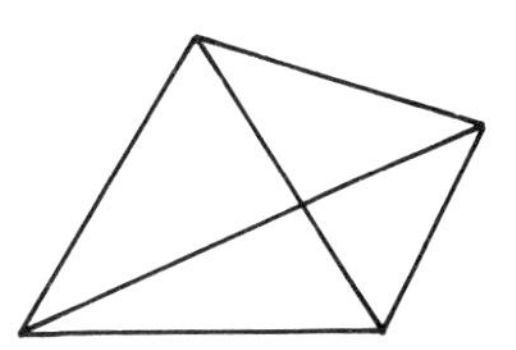

We show the cycle types of these symmetries, considered as permutations of the vertices, and the number of elements fixed by each symmetry, in the table below.

Symmetry g	*Cycle type*	$\#(\mathit{Fix}(g))$
r_0	x_1^8	c^8
r_1, r_3, r_5, r_7	x_8	c
r_2, r_6	x_4^2	c^2
r_4	x_2^4	c^4
s_1, s_2, s_3, s_4	$x_1^2x_2^3$	c^5
t_1, t_2, t_3, t_4	x_2^4	c^4

It follows from Burnside's Theorem that the number of distinct patterns is

$$\tfrac{1}{16}(c^8 + 4c^5 + 5c^4 + 2c^2 + 4c).$$

10.2.3. In the following table you will find the symmetries of the square, and the number of colourings of the 5×5 chessboard fixed by each of them.

g	e	a	b	c	p	q	r	s
$\#(Fix(g))$	3^{25}	3^7	3^{13}	3^7	3^{15}	3^{15}	3^{15}	3^{15}

Hence, by Burnside's Theorem, the number of different patterns is

$$\tfrac{1}{8}(3^{25} + 4(3^{15}) + 3^{13} + 2(3^7)) = 105\,918\,450\,471.$$

10.2.4. The symmetries of the figure are the same as the symmetries of the square. We list these symmetries, the cycle type of the corresponding permutation of the 45 small squares which make up the figure, and the number of colourings, using three colours, which these symmetries fix.

Symmetry g	*Cycle type*	$\#(Fix(g))$
e	x_1^{45}	3^{45}
$a,\ c$	$x_1x_4^{11}$	3^{12}
b	$x_1x_2^{22}$	3^{23}
$p,\ q$	$x_1^9x_2^{18}$	3^{27}
$r,\ s$	$x_1^3x_2^{21}$	3^{24}

Hence, by Burnside's Theorem, the number of different patterns is

$$\tfrac{1}{8}(3^{45} + 2(3^{27}) + 2(3^{24}) + 3^{23} + 2(3^{12}))$$
$$= 369\,289\,090\,307\,628\,997\,911.$$

10.2.5. The rotational symmetries of a cube were listed in the solution to Exercise 7.3.3. We saw that there were 24 symmetries in all, of five different kinds. We list them again in the following table, together with the cycle types of the associated permutations of the faces of the cube, and the number of colourings fixed by them.

Kind of rotation	*Number of this kind*	*Cycle type*	$\#(Fix(g))$
Identity	1	x_1^6	3^6
Rotations through $\pi/2$ about axes joining midpoints of opposite faces	6	$x_1^2x_4$	3^3

Kind of rotation	*Number of this kind*	*Cycle type*	*#(Fix(g))*
Rotations through π about axes joining midpoints of opposite faces	3	$x_1^2x_2^2$	3^4
Rotations through π about axes joining midpoints of opposite edges	6	x_2^3	3^3
Rotations through $2\pi/3$ about axes joining opposite vertices	8	x_3^2	3^2

It therefore follows from Burnside's Theorem that the number of distinct colourings of the cube, using the three colours red, white and blue, is

$$\tfrac{1}{24}(3^6 + 3(3^4) + 12(3^3) + 8(3^2)) = \tfrac{1}{24}(1368) = 57,$$

a result which has been attributed to H. J. Heinz.

10.2.6. Here we let the group G of the rotational symmetries of a cube act on the set X of cubes with the numbers 1 to 6 on their faces. These six numbers can be arranged on the faces of a cube in 6! ways. The identity symmetry fixes each of these arrangements, but every other symmetry changes the arrangement. Thus $\#(\mathrm{Fix}(e)) = 6!$, and for every other symmetry g, $\#(\mathrm{Fix}(g)) = 0$. Hence, by Burnside's Theorem, the number of different arrangements is

$$\frac{6!}{24} = 30.$$

10.2.7. For each solid we list the rotational symmetries, the cycle types of the corresponding permutations of the faces, and hence the number of colourings which they fix.

First, the regular octahedron:

Kind of rotation	*Number of this kind*	*Cycle type*	*#(Fix(g))*
Identity	1	x_1^8	c^8
Rotations through $2\pi/3$ about axes joining midpoints of opposite faces	8	$x_1^2x_3^2$	c^4
Rotations through π about axes joining midpoints of opposite edges	6	x_2^4	c^4

Kind of rotation	*Number of this kind*	*Cycle type*	*#(Fix(g))*
Rotations through $\pi/2$ about axes joining opposite vertices	6	x_4^2	c^2
Rotations through π about axes joining opposite vertices	3	x_2^4	c^4

It follows from Burnside's Theorem that the number of different colourings is

$$\tfrac{1}{24}(c^8 + 17c^4 + 6c^2).$$

Next, the regular dodecahedron:

Kind of rotation	*Number of this kind*	*Cycle type*	*#(Fix(g))*
Identity	1	x_1^{12}	c^{12}
Rotations through $2\pi/5$ about axes joining midpoints of opposite faces	24	$x_1^2x_5^2$	c^4
Rotations through π about axes joining midpoints of opposite edges	15	x_2^6	c^6
Rotations through $2\pi/3$ about axes joining opposite vertices	20	x_3^4	c^4

It follows from Burnside's Theorem that the number of distinct colourings is

$$\tfrac{1}{60}(c^{12} + 15c^6 + 44c^4).$$

Finally, the regular icosahedron:

Kind of rotation	*Number of this kind*	*Cycle type*	*#(Fix(g))*
Identity	1	x_1^{20}	c^{20}
Rotations through $2\pi/3$ about axes joining midpoints of opposite faces	20	$x_1^2x_3^6$	c^8
Rotations through π about axes joining midpoints of opposite edges	15	x_2^{10}	c^{10}
Rotations through $2\pi/5$ about axes joining opposite vertices	24	x_5^4	c^4

It follows that the number of different colourings is

$$\tfrac{1}{60}(c^{20} + 15c^{10} + 20c^{8} + 24c^{4}).$$

CHAPTER 11

11.2.1. We assign the novels by Jane Austen, Charles Dickens and Anthony Trollope the weights a, d and t respectively. The store enumerator is $(6a + 14d + 48t)$, and hence the inventory of all the possible choices of the students is

$$(6a + 14d + 48t)^6.$$

We want the coefficient of $a^2d^2t^2$, and by the multinomial theorem this is

$$\frac{6!}{2!2!2!} \times 6^2 14^2 48^2 = 1\,463\,132\,160.$$

11.2.2. We saw, in Chapter 8.1 that there are six different patterns. The corresponding pattern inventory is

$$w^4 + w^3b + 2w^2b^2 + wb^3 + b^4.$$

11.3.1. The cycle index of S_5 is

$$\tfrac{1}{120}(x_1^5 + 10x_1^3x_2 + 15x_1x_2^2 + 20x_1^2x_3 + 20x_2x_3 + 30x_1x_4 + 24x_5).$$

The cycle index of S_6 is

$$\tfrac{1}{720}(x_1^6 + 15x_1^4x_2 + 45x_1^2x_2^2 + 15x_2^3 + 40x_1^3x_3 + 120x_1x_2x_3$$
$$+ 40x_3^2 + 90x_1^2x_4 + 90x_2x_4 + 144x_1x_5 + 120x_6).$$

11.3.2. (a) $\tfrac{1}{8}(x_1^4 + 3x_2^2 + 2x_1^2x_2 + 2x_4)$.

(b) $\tfrac{1}{12}(x_1^6 + 4x_2^3 + 3x_1^2x_2^2 + 2x_3^2 + 2x_6)$.

(c) $\tfrac{1}{24}(x_1^6 + 6x_2^3 + 3x_1^2x_2^2 + 6x_1^2x_4 + 8x_3^2)$.

11.4.1 We list the symmetries of a square and the cycle types of the corresponding permutation of the 25 squares of the 5×5 chessboard:

e	a	b	c	p	q	r	s
x_1^{25}	$x_1x_4^6$	$x_1x_2^{12}$	$x_1x_4^6$	$x_1^5x_2^{10}$	$x_1^5x_2^{10}$	$x_1^5x_2^{10}$	$x_1^5x_2^{10}$

It follows that the cycle index of the group is

$$\tfrac{1}{8}(x_1^{25} + 2x_1x_4^6 + x_1x_2^{12} + 4x_1^5x_2^{10}).$$

Therefore the pattern inventory is

$$\tfrac{1}{8}((b + w)^{25} + 2(b + w)(b^4 + w^4)^6 + (b + w)(b^2 + w^2)^{12} + 4(b + w)^5(b^2 + w^2)^{10}).$$

We seek the coefficient of $b^{15}w^{10}$. The coefficient of $b^{15}w^{10}$ in $(b + w)^{25}$ is $C(25, 15) = 3\,268\,760$. There is no term involving $b^{15}w^{10}$ in $2(b + w)(b^4 + w^4)^6$. The term involving $b^{15}w^{10}$ in $(b + w)(b^2 + w^2)^{12}$ comes by multiplying the b from $(b + w)$ by $b^{14}w^{10}$ from $(b^2 + w^2)^{12}$, and so its coefficient is $C(12, 7) = 792$. The term involving $b^{15}w^{10}$ from $(b + w)^5(b^2 + w^2)^{10}$ arises in the following way. The product of b^5 from $(b + w)^5$ and $b^{10}w^{10}$ from $(b^2 + w^2)^{10}$, which has coefficient $C(5, 5) \times C(10, 5) = 252$. The product of b^3w^2 from $(b + w)^5$ and $b^{12}w^8$ from $(b^2 + w^2)^{10}$, which has coefficient $C(5, 3) \times C(10, 6) = 2100$. The product of bw^4 from $(b + w)^5$ and $b^{14}w^6$ from $(b^2 + w^2)^{10}$, which has coefficient $C(5, 1) \times C(10, 7) = 600$. Hence the coefficient of $b^{15}w^{10}$ in $4(b + w)^5(b^2 + w^2)^{10}$ is

$$4(252 + 2100 + 600) = 11\,808.$$

It follows that the coefficient of $b^{15}w^{10}$ in the pattern inventory is

$$\tfrac{1}{8}(3\,268\,760 + 0 + 792 + 11\,808) = 410\,170.$$

This gives the number of different patterns with 15 black squares and ten white squares.

11.4.2. The cycle index of the group of permutations of the eight small triangles in the figure, corresponding to the symmetries of the square, is

$$\tfrac{1}{8}(x_1^8 + 5x_2^4 + 2x_4^2)$$

and hence the pattern inventory for colourings using red, white and blue, to which we assign the weights, r, w, b, respectively, is

$$\tfrac{1}{8}((r + w + b)^8 + 5(r^2 + w^2 + b^2)^4 + 2(r^4 + w^4 + b^4)^2).$$

We seek the coefficient of $r^2w^2b^4$ in this pattern inventory. This coefficient is

$$\frac{1}{8}\left(\frac{8!}{2!2!4!} + 5 \times \frac{4!}{1!1!2!} + 2 \times 0\right) = \tfrac{1}{8}(420 + 60) = 60.$$

11.4.3. We calculated the relevant cycle index in answering Exercise 11.3.2(c), when we saw that it is

$$\tfrac{1}{24}(x_1^6 + 6x_2^3 + 3x_1^2x_2^2 + 6x_1^2x_4 + 8x_3^2).$$

The pattern inventory of colourings using red, white and blue is thus

$$\tfrac{1}{24}((r + w + b)^6 + 6(r^2 + w^2 + b^2)^3 + 3(r + w + b)^2(r + w^2 + b^2)^2 + 6(r + w + b)^2(r^4 + w^4 + b^4) + 8(r^3 + w^3 + b^3)^2).$$

We seek the coefficient of r^3w^2b in this pattern inventory. It can easily be seen that only the expressions $(r + w + b)^6$ and $3(r + w + b)^2(r^2 + w^2 + b^2)$ in the pattern inventory yield terms involving r^3w^2b. Thus the coefficient of r^3w^2b is

$$\frac{1}{24}\left(\frac{6!}{3!2!1!} + 3 \times 2 \times 2\right) = 3.$$

Hence there are three different colourings of the cube with three red faces, two white faces and one blue face.

11.4.4. The cycle index of the group of permutations of the faces of a regular octahedron, corresponding to the rotational symmetries, can be deduced from the answer to Exercise 10.2.7. It is

$$\tfrac{1}{24}(x_1^8 + 8x_1^2x_3^2 + 9x_2^4 + 6x_4^2).$$

Hence the pattern inventory of the colourings using red, white and blue is

$$\tfrac{1}{24}((r + w + b)^8 + 8(r + w + b)^2(r^3 + w^3 + b^3)^2 + 9(r^2 + w^2 + b^2)^4 + 6(r^4 + w^4 + b^4)^2).$$

It follows that the coefficient of $r^4w^2b^2$ is

$$\tfrac{1}{24}(420 + 0 + 108 + 0) = 22.$$

11.4.5. A position in the game of noughts and crosses consists of an arrangement of noughts and crosses on a 3×3 grid. For example

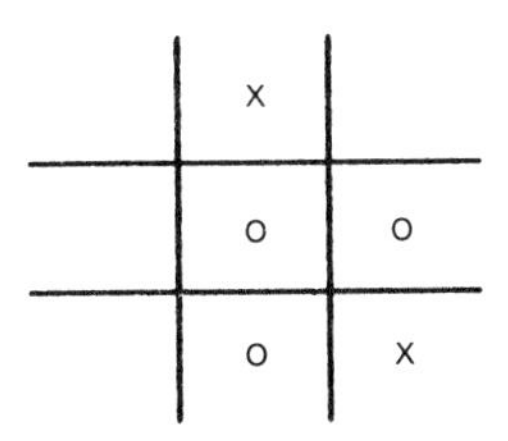

These positions correspond to the colourings of a 3×3 chessboard using three colours and we can thus use the machinery of Pólya's Theorem to count them. The rules of noughts and

crosses impose several restrictions on which positions can arise, as follows.

(a) Assuming that 'noughts' goes first, in any position there are either equal numbers of noughts and crosses, or there is one more nought than cross.
(b) When there is a line of noughts, the game must have ended when this was created with a move by 'noughts', and so there is one more nought than cross.
(c) When there is a line of crosses the game must have ended with a move by 'crosses' so there is an equal number of nought and crosses.

Pólya's Theorem is designed to count the number of different positions which meet requirement (a). We will worry about the other requirements later. The cycle index of the group of permutations corresponding to the symmetries of the noughts and crosses grid is

$$\tfrac{1}{8}(x_1^9 + 2x_1x_4^2 + x_1x_2^4 + 4x_1^3x_2^3).$$

If we assign the weights a, b and c to noughts, crosses and blank squares respectively, then the pattern inventory is

$$\tfrac{1}{8}((a+b+c)^9 + 2(a+b+c)(a^4+b^4+c^4)^2 + (a+b+c)(a^2+b^2+c^2)^4 + 4(a+b+c)^3(a^2+b^2+c^2)^3).$$

Positions corresponding to requirement correspond to terms in the pattern inventory in which the coefficients of a and b are equal, or the coefficient of a exceeds the coefficient of b by 1. The terms which satisfy this condition and their coefficients are given in the following table:

c^9	1
ac^8	3
abc^7	12
a^2bc^6	38
$a^2b^2c^5$	108
$a^3b^2c^4$	174
$a^3b^3c^3$	228
$a^4b^3c^2$	174
a^4b^4c	89
a^5b^4	23
Total	850

The only way I can think of to count the positions which although meeting condition (a) do not meet conditions (b) and (c) is just to list them. According to my reckoning there are 83 of these positions. It follows that there are $850 - 83 = 767$ different positions that can arise in a game of noughts and crosses.

11.6.1. We list the cycle types of the elements of S_4^* in the following table. The cycle types of the elements of S_4 and the number of elements of each type were given in section 11.3. So all there remains to do is to calculate the cycle types of the corresponding elements of S_4^*.

Cycle type of $\pi \in S_4$	*No. of permutations of this cycle type*	*Cycle type of* $\pi^* \in S_4^*$
x_1^4	1	x_1^6
$x_1^2x_2$	6	$x_1^2x_2^2$
x_2^2	3	$x_1^2x_2^2$
x_1x_3	8	x_3^2
x_4	6	x_2x_4

It follows that the pattern inventory is

$$\tfrac{1}{24}((1 + c)^6 + 9(1 + c)^2(1 + c^2)^2$$
$$+ 8(1 + c^3)^2 + 6(1 + c^2)(1 + c^4))$$
$$= 1 + c + 2c^2 + 3c^3 + 2c^2 + c^5 + c^6.$$

We have already listed, in the solution to Exercise 9.1.1, all the graphs with four vertices. This pattern inventory, whose coefficients add up to 11, confirms that our list did include all the different graphs with four vertices.

11.6.2. The cycle index of S_6 is given in the solution to Exercise 11.3.1. We need to calculate the cycle types of the corresponding elements of S_6^*. These are given in the following table:

Cycle type of $\pi \in S_6$	*No. of permutations of this cycle type*	*Cycle type of* $\pi^* \in S_6^*$
x_1^6	1	x_1^{15}
$x_1^4x_2$	15	$x_1^7x_2^4$
$x_1^2x_2^2$	45	$x_1^3x_2^6$
x_2^3	15	$x_1^3x_2^6$

Cycle type of $\pi \in S_6$	No. of permutations of this cycle type	Cycle type of $\pi^* \in S_6^*$
$x_1^3x_3$	40	$x_1^3x_3^4$
$x_1x_2x_3$	120	$x_1x_2x_3^2x_6$
x_3^2	40	x_3^5
$x_1^2x_4$	90	$x_1x_2x_4^3$
x_2x_4	90	$x_1x_2x_4^3$
x_1x_5	144	x_5^3
x_6	120	$x_3x_6^2$

It follows that the pattern inventory is

$$\tfrac{1}{720}((1 + c)^{15} + 15(1 + c)^7(1 + c^2)^4 + 60(1 + c)^3(1 + c^2)^6$$
$$+ 40(1 + c)^3(1 + c^3)^4 + 120(1 + c)(1 + c^2)(1 + c^3)^2(1 + c^6)$$
$$+ 40(1 + c^3)^5 + 180(1 + c)(1 + c^2)(1 + c^4)^3$$
$$+ 144(1 + c^5)^3 + 120(1 + c^3)(1 + c^6)^2)$$
$$= 1 + c + 2c^2 + 5c^3 + 9c^4 + 15c^5 + 21c^6 + 24c^7 + 24c^8$$
$$+ 21c^9 + 15c^{10} + 9c^{11} + 5c^{12} + 2c^{13} + c^{14} + c^{15}.$$

The sum of the coefficients in this polynomial is 156. This shows that there are 156 different graphs with six vertices. We can read off from the different terms how many different graphs there are with a given number of edges. So, for example, there are 21 different graphs with six vertices and six edges. If you try to list these, you will appreciate the power of Pólya's Theorem.

11.6.3.

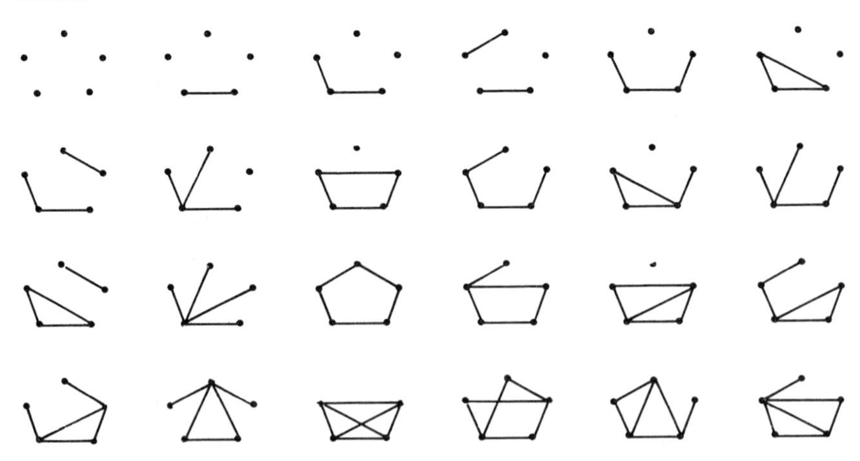

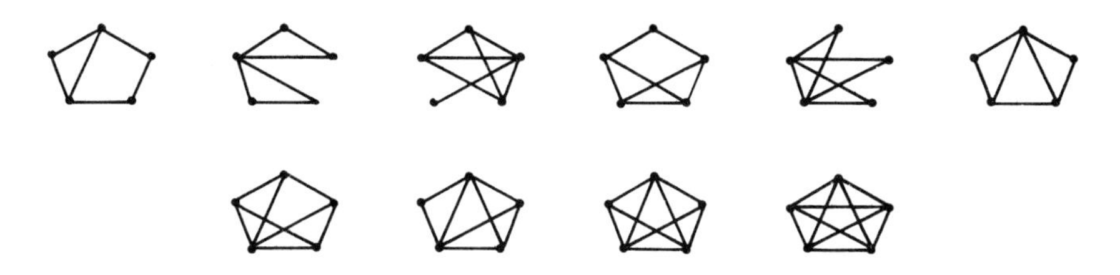

11.6.4. Suppose that π_1 and π_2 are permutations from S_n which have the same cycle type. It follows that they are conjugate permutations. That is there is some permutation $\sigma \in S_n$ such that $\pi_2 = \sigma\pi_1\sigma^{-1}$. (If this piece of group theory is not familiar to you, note that if

$$\pi_1 = (n_1 n_2 \ldots n_k) \ldots$$

and

$$\pi_2 = (n_1' \, n_2' \ldots n_k') \ldots$$

then it is sufficient to take σ to be the permutation

$$\begin{pmatrix} n_1 & n_2 & \ldots & n_k \\ n_1' & n_2' & \ldots & n_k' \end{pmatrix} \ldots)$$

It is easily checked (see the solution to Exercise 11.6.6) that it follows that

$$\pi_2^* = \sigma^* \pi_1^* \sigma^{*-1}$$

and hence that π_1^* is conjugate to π_2^* in S_n^*. It follows that π_1^* and π_2^* have the same cycle type in S_n^*.

11.6.5. No. We have already seen several counterexamples. For example, in the solution to Exercise 11.6.1 we saw that permutations in S_4 of cycle types x_1x_2 and x_2^2 both correspond to permutations in S_4^* of cycle type $x_1^2x_2^2$.

11.6.6. We show first that the mapping $\Phi : S_n \to S_n^*$ defined by $\Phi(\pi) = \pi^*$ is a homomorphism. Suppose $\pi_1, \pi_2 \in S_n$ and that $1 \leqslant i < j \leqslant n$. Then

$$\begin{aligned} (\pi_1\pi_2)^*(\{i, j\}) &= \{\pi_1\pi_2(i), \pi_1\pi_2(j)\} \\ &= \{\pi_1(\pi_2(i)), \pi_1(\pi_2(j))\} \\ &= \pi_1^*(\{\pi_2(i), (\pi_2(j)\}) \\ &= \pi_1^*(\pi_2^*(\{i, j\})) \\ &= \pi_1^*\pi_2^*(\{i, j\}), \end{aligned}$$

and hence $(\pi_1\pi_2)^* = \pi_1^*\pi_2^*$, showing that Φ is a homomorphism. It remains only to show that Φ is injective (one-to-one). Suppose, then, that $\pi_1 \neq \pi_2$. Then for some i, $\pi_1(i) \neq \pi_2(i)$. Choose $j \in \{1, \ldots, n\}$ so that $\pi_1(j)$ is distinct from $\pi_1(i)$ and $\pi_2(i)$. This is possible provided that $n > 2$. Then

$$\begin{aligned} \pi_1^*(\{i, j\}) &= \{\pi_1(i), \pi_1(j)\} \\ &\neq \{\pi_2(i), \pi_2(j)\} = \pi_2^*(\{i, j\}), \end{aligned}$$

as neither $\pi_1(i)$ nor $\pi_1(j)$ is equal to $\pi_2(i)$. Thus $\pi_1^* \neq \pi_2^*$, showing that Φ is injective. This completes the proof that Φ is an isomorphism.

Suggestions for further reading

I hope that you have ended your reading of this book wanting more. So what follows is a short list of books where you can read more about the topics in this book, and about aspects of combinatorics that I have not mentioned at all. It is by no means a complete list of books in this area, but merely a personal selection.

For alternative introductions to combinatorics at the same level of this book you should read either the treatment by Anderson [2] or that by Cohen [4]. Combinatorics often appears under the title of 'discrete mathematics' and there are now many books with these words in their title. Many of these cover topics which form the mathematical basis of computer science such as elementary set theory, and logic. A very distinguished exception is Biggs's work [3], which is an excellent introduction to genuine combinatorics.

When you feel ready to tackle more advanced books on combinatorics, you should try one of three works listed below, by Goulden and Jackson [5], Krishnamurthy [7] or Liu [9]. If you enjoy formulas, then John Riordan's book is the one for you.

There are lots of introductions to group theory. My first suggestion is Reg Allenby's book [1], partly because he is a colleague of mine in the School of Mathematics at Leeds University, but mainly because it is extremely well written. If you want a book that just covers groups, then my suggestion would be that by W. Ledermann. There are not so many introductions to graph theory, but in any case I see no reason to look beyond Robin Wilson's book [13], whose merits are indicated by its now being in its third edition.

Pólya's own approach to combinatorics can be found in Pólya, Tarjan and Woods's book [10], and the paper in which his theorem was first published is in the work by Pólya and Read [11], which I have already mentioned earlier in this text. To find out more about the specific problem of counting graphs of different types you should consult Harary and Palmer's work [6].

[1] R. B. J. T. Allenby (1983) *Rings, Fields and Groups,* London.
[2] Ian Anderson (1989) *A First Course in Combinatorial Theory,* 2nd edition, Oxford.
[3] Norman Biggs (1985) *Discrete Mathematics,* Oxford.
[4] Daniel I. A. Cohen (1978) *Basic Techniques of Combinatorial Theory,* New York.
[5] I. P. Goulden and D. M. Jackson (1983) *Combinatorial Enumeration,* New York.
[6] Frank Harary and Edgar M. Palmer (1973) *Graphical Enumeration,* New York and London.
[7] V. Krishnamurthy (1986) *Combinatorics: Theory and Applications,* Chichester.
[8] W. Ledermann (1973) *Introduction to Group Theory,* Harlow.
[9] Chung Laung Liu (1968) *Introduction to Combinatorial Mathematics*, New York.
[10] George Pólya, Robert E. Tarjan and Donald R. Woods (1983) *Notes on Introductory Combinatorics,* Cambridge, Massachusetts.
[11] George Pólya and R. C. Read (1987) *Cominatorial Enumeration of Groups, Graphs, and Chemical Compounds,* New York.
[12] John Riordan (1968) *Combinatorial Identities*.
[13] Robin J. Wilson (1985) *Introduction to Graph Theory,* 3rd edition, Harlow.

List of special symbols

D_n	187
π^*	188
S_n^*	188

Index